from Field & Forest

from Field & Forest

An artist's year
in paint and pen

Anna Koska

PAVILION

For Mum and Dad

First published in the United Kingdom in 2021 by Pavilion
43 Great Ormond Street
London
WC1N 3HZ

ISBN 9781911641766

A CIP catalogue record for this book is available from the British
Library.

10 9 8 7 6 5 4 3 2 1

Reproduction by Rival Colour Ltd, UK
Printed and bound by 1010 Printing International Ltd, China

www.pavilionbooks.com

Autumn 6

Winter 42

Spring 72

Summer 102

Autumn

Oakmoss Lichen

Plenty

We moved to this patch of wild some 17 years ago.

The dream was to let our children run feral. Our previous home had been up four flights of stairs in an apartment block on the side of a busy road. It was perfectly fine and we'd managed okay with the first two children, but it had become more and more obvious that this wasn't ideal for the kind of family we'd hoped to grow.

So taking a fairly giant leap, we threw everything we had (and didn't have, to be honest) at the opportunity to give the children, and us, a different kind of life.

Our house sits on the edge of a field that is surrounded by woods and water. We have the joy of watching all of nature's grandest performances. And although she repeats herself (every year!) no two shows have been the same.

Pretty soon after emptying boxes, filling shelves and cupboards to overflowing (and lugging the remaining unopened boxes up to the attic) we began to realise the less than romantic practicality of keeping our new home warm; we'd discovered just how utterly rubbish the central heating was. There was an awful lot of hissing, wining and general bellyaching for very little output. Not a dissimilar noise to the one the plumber made when we asked how much it would cost to get it fixed... much sucking of teeth! The warmest room was the little space beneath the stairs where there's a loo, a spreading mass of boots and shoes, too many coats and reusable shopping bags, and a washing machine. I'm sure that each one of us has, at some point, wished that this was our bedroom!

I was brought up in a very old cottage in Cornwall, where we endured the same miserable challenge. With granite stone walls as broad as a cow, we enjoyed the cool in the summers but winters could be quite brutal. I've memories of spreading out my school uniform along the front bar of my parents' Rayburn in the evening. And, so long as we'd remembered to stoke the stove, I'd at least be able to run down to the kitchen first thing and put on warm clothes.

I would wake up in a crazed muddle of sheets and frost-damp blankets and admire the feathered swirls of ice art on the inside of my bedroom window. Bracing myself, I'd perform a mental rocket-launch countdown to force myself out of bed and down the stairs. The race was made all the more worthwhile since the first one down also got to have the cream off the top of the (glass) bottle of milk.

Standing in a shivering stoop, two pairs of socks would be pulled up to each knee, always the vest beneath the shirt and, gripping the cuffs firmly, I'd pull on the regulation nylon jumper. School rules dictated that the length of a skirt wasn't allowed to creep above the knees. But, of course, we all wanted a shorter skirt. Regardless of the endless, nagging assault to which I subjected my poor mother throughout summer, come winter I was quietly relieved that my skirt was still that little bit longer. All fond memories now, but the phrase 'rising damp' will forever trigger an involuntarily shudder from my kidneys.

We're lucky enough to have a bit of woodland here. But the previous owner had been all about the flower beds and perfect parallel stripes on the lawn. So the woods, meanwhile, had been left to 'manage' themselves, which meant that we had a ready supply of fallen and sickly trees to use. And so we quickly realised that in order to stay warm we'd need to keep home fires burning – that, and wear more clothing.

As the children grew older they were able to help more with the sawing, cutting and stacking and so it became very much a team effort. Much soup, warm bread and cake, and the odd hip flask of whisky, made this chore far more appealing for all concerned. And the irony is that throughout this cheerful production line of activity we'd inevitably begin to peel off layers as cheeks became ruddy and our core temperatures rose to the point of feeling like the advert for Ready Brek.

But there's something undeniably right and proper about this. The joy of burning logs shouldn't come free. There needs to be an exchange of some sort. By clearing away the dead and dark areas, not only were we rewarded with fuel to keep our home warm, but also this small acreage of forest has begun to wake up and thrive.

Young trees that had become light-starved and etiolated are now standing tall and stretching out, as well as up. Lower canopies have begun to spread, perfect for pheasant,

rabbit and deer, and a little pond we'd dug out is now
flourishing with the increased sunlight – come spring it is
teeming with life.

It's 7.30 at night. Thrusting socked feet into cold boots
and shrugging into the nearest coat, I've a head torch on
and with every step its saggy elastic keeps slipping down to
perch the light on the bridge of my nose. A bit annoying,
but it's either this or barrowing out into the inky night
with a fair chance of ending up in the bank of broom. I'm
making my way down to the log store my husband has
built. It's large by many folk's standards; it could even be
classed as greedy, unnecessary. But the truth is, we will
work our way through at least three of its bays by March
next year.

And now I'm here, empty barrow ready to fill. I'm
captivated by the myriad colours and textures in this
unintended collage of log ends, stacked and packed tight,
ready for service. From softest ochre to clotted cream,
deepest amber to summer blonde; depending on the age
and health of the tree, and the time it was cut, each clutch
of logs will gradually become a gentler palette of hues.
Added to this, there's the exquisite lacework of lichens
that have decided to make this particular stack, an oak,
their home. They stretch out as a cloak, at times a flat and
fleshy carpet of leaf-like growth, while other logs proudly
sport a delicate explosion of fronds that put me in mind of
my A-level Biology studies, the revelation that the internal
network of bronchial tubes bears more than just a passing

resemblance to the skyward stretching spread of arms on a tree. Lungs, all three.

I slip into the familiar rhythm of grab, swing and drop and the barrow fills.

We've never been particularly focused on material wealth, but this is perhaps one of the most comforting sights I know. As we gallop towards winter, this view makes me as happy as a full fridge might, or seeing all the jams, preserves and honey gathered and ready in the larder.

And I feel genuinely rich in the certainty of warmth, whatever this winter decides to do.

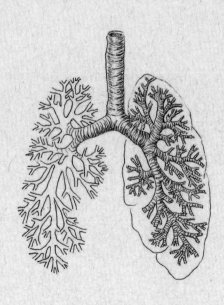

Mallow

Spinning Classes

I'm writing this having washed myself down from chin to toe, thus saving the keyboard from developing any further qwerty idiosyncrasies.

With glorious hindsight, it may have been more efficient to have brought in a jet spray and just stood in the middle of the kitchen, while some kind soul hosed me down, along with the counters, walls, floor and the few brave wasps that insisted on hanging around, such was the degree of stickiness.

But I really didn't care. The whole palaver was genuinely bloody fabulous. As it is every year.

Come the beginning of September the bees and I have an understanding that I will, with the upmost care – and as swiftly as possible to cause the least stress – remove just one of the 'supers' of honey from each hive.

'Super' is the name for the box that contains frames holding strips of beeswax. It sits above the 'nursery' or brood box. From early spring the bees will begin to plan for the many new youngsters that their heroic queen will be producing. For a strong, healthy colony numbers can swell from around 10,000 bees in early spring to around 80,000 in peak summer. That's a lot of mouths to feed. The bees will busy themselves adding to and sculpting the wax sheets into the most beautiful, uniform wall of slightly upward tilting cells, ready to receive the gathered nectar.

If you're fortunate enough to get close to an open hive, the smell of freshly shaped wax is intoxicating.

And so the business of filling the larder begins. Each bee will bring foraged nectar and transfer it to another bee. This bee will do the same, and so a chain of bees will continue this deceptively simple game of pass-the-nectar until there's enough digestive enzymes contained within the liquid to help it convert to the golden liquid we recognise as honey. As the frames fill up, the bees will work out a rota for fanning the nectar with their wings, to remove the excess moisture. Once quality control has ticked off this stage of honey production, each of these little cells will be capped with a fresh lid of wax, sealing it off to finish its conversion to honey and keep it safe until needed by the colony. If this was a little more information than you'd planned on knowing, and indeed it's confirmed a long-held suspicion that honey really is 'bee spit', please accept my apologies. It is, sort of, but it's a much more complex and clever procedure than you might imagine. And the end result is a gift, truly and utterly priceless.

If it's been a good season for them with a bountiful supply of blooms to forage among, a full and busy colony of bees can fill upwards of 24 frames this way. That's approximately 90lb of honey. And my enduring promise to 'my' bees is that I will always leave them with enough to take them through the winter and into spring.

I heated a bread knife on the Aga hot plate… the smell of singed wax as I shallowly sliced the capping from the comb... the tsunami of perfume as the honey unfolded itself from the cells... mallow, broom and clover... the summer reanimated.

I carefully placed the uncapped frames in the mechanical honey extractor I'd hired and turned the handle. Momentum gathered and honey began to fly from the outer cells. The honey extractor, essentially a big plastic tub, began to shudder and attempt an enthusiastic twerk across my kitchen floor. I braced it against a cupboard with my knees and cranked up the speed, entranced by the flickering little arcs of amber sweetness as they hit the tub wall and began an unhurried descent to pool at the bottom. Wow. Miraculous.

Stickiness didn't really arrive until I needed to filter this honey. I found that by using a child's high chair (a Tripp Trapp in case you're planning to have a go) I could position the spinner on top and place a smaller tub, with sieve, below with 'just' enough overlap to avoid floor puddles. This worked(ish) until I needed to get the last bits out. I took the high chair out of the equation and, clasping the tilted extractor between my knees, I reached through the internal workings to the persistent little puddle at the bottom, and began scooping it towards the tapped exit. I got the last of the honey!

I also got honey all over me... my hands, wrists, elbows, arms, neck, chin, hair, and smile.

Take pot of honey to post office for Julie!

Every year it's the same routine. And every year I feel the familiar flood of happy awe.

I've now filled and lidded a dozen pots and placed them on a shelf in the larder… little glass jars of sunshine for winter's grey, ready to give to friends. The irony's not lost on me…

Suiting up again I head back down to the hives to check they've settled down after my rude interruption to their day. I sit down on the edge of the hive stand and, placing my gloved hands in my lap, I just watch for a while. It's about five in the afternoon, and this hive is now a little in the shade cast by the old storage shed behind me, but the scent of sun-warmed cedar still dominates the air. The level of activity on the bees' doorstep is still quite high and will continue until the light fades and the temperature drops below shirt-sleeves comfortable. A pollen-heavy bee has just bumped onto the step. It's not the most elegant of landings. Her wings are a little frayed at the edges; there's no doubt it's been a long summer for this one. Another bee, just about to launch, instead turns around and heads back to this exhausted forager. She touches her face, caresses it even, seems to be offering support as she endeavours to make her way up the ramp and into the hive. Little moments like these would be so easy to anthropomorphise. But I resist the temptation to make it cute or more tender than it already clearly is. Call it practical, survival of the super-organism, downplay it as much as possible, but the truth of it is that these bees continue to amaze me with

their open displays of patience and care of one another;
it can make us feel very insignificant and foolish for the
demands we make on how we want our lives to be, what
we expect the world to give us. For these little ones, their
only mission is to do everything possible to nurture one
another, and in so doing secure the survival of their colony
through to the next year.

The hives will begin to wind down for winter now. Each
queen will already have slowed down her egg production,
the last few thousand being a special type of bee, which
will be a bit plumper with extra reserves of fat and protein
to help them survive the colder months. The whole super
organism will set about conserving energy, building and
saving stores, and guard bees will continue to lurk at their
narrowed door, spiked bottoms waggling at potential
wasp intruders who would wipe out an entire winter store
in a heartbeat.

I thank them for their honey and wish them well for the
coming season.

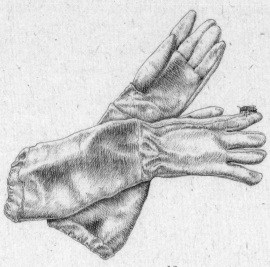

The Millpond

Down here, at the bottom of our field, near the gentle
ripple of water, there's still a little light left but it's a time
when the edges of the day become softened. I'm standing
beneath one of the three field maples we planted 15 years
ago, and I can see her arms are laden with crowds of
winged seeds, waiting in clusters like a fleet of butterflies,
unfurled and flushing to pink, readying for autumn winds.
And by all accounts they're on their way. Yet the air still
holds the vestigial warmth of summer's touch.

Most of my meanders through this green and tufted
ocean are at the very beginning of the day or just as the
last of the light dips below the crown of oak and hornbeam
that lies to the west, beyond our house. Tonight's sky is
flooded with colour, from soft blush to deepest rouge, a
bruise slowly spreading.

We've lived on the edge of this field long enough for me
to know her every dip and curve as she makes her way
down towards the millpond at the bottom. It's not our
pond, but we get to see this huge mirror reflect every mood

Goose eye colours:

Canada ~ brown
Greylag ~ deep brown
Domestic ~ blue

of every season, and we've become an enthralled audience to the wild and feathered theatre it stages. With regulars, interlopers and the occasional appearance of random strangers, it plays out like a version of some small-town Netflix series... dramatic cliff-hangers, petty squabbles and romance included.

This evening, it's the turn of the laughing ducks and the rippled echo-call of the coot. As the sun continues her journey west, the Canada geese swing round and down along the aisle of old pine to take up their moorings. As they near the water, they stop calling, fully focused on the business of docking. Standing close by, I can hear the wind grazing at their wings, vibrating primary feathers. Both a whisper and a moan, it's a melding of two pitches to create a unique sound, polyphonic almost. Their feet lower to plough great folds of water. There's an outburst of fresh calls from those already settled, the new arrivals answer, and so the conversation flows until, after a while, all becomes quiet once more.

As autumn takes hold and the trees become gilded, with petticoats loosening, the greylag geese will return, which usually enriches the plot somewhat... many mutterings and much griping over mooring lines and mates usually ensues. But for now, the current cast seems content with the slowing pace. With families raised and predators less pressing, it's easy to imagine that they're all now in need of a little downtime as they recover from what was undoubtedly a stressful summer of child-rearing.

Raised by parents who bred and grew most of our food, goose did occasionally feature on the menu during Christmas or Easter celebrations. But over time my father fell in love with these proud and fiercely loyal birds, and so became less inclined to keep them for food. I have the happiest of memories of walking across a daisy-dotted lawn with a chirpy trail of young geese hurrying after me, small stubs-for-wings extended for balance. Like everyone in my family, I became used to translating the different calls and mutterings they made. They were very much part of our lives, often appearing in the kitchen to chatter, ask for a hunk of bread, or indeed recuperate when nursing was needed. The Rayburn door was often left ajar to warm a tired chick, or quite regularly hatch an egg. Dad still keeps geese, and I can't imagine a time when he won't.

Since living here, I've grown familiar once more with the general chatter among the pond's frequent visitors. A call to clear the runway for take off and landing is quite different from a warning to back away, you're standing/floating a little too close for my comfort. The call to those in the vicinity that a stranger is near is piercing and will usually culminate in a Mexican wave of high-pitched shrieks, the contagion of fear spreading rapidly, to enable all in the community to be on guard. Often the ducks take up the baton and run with it, adding their harsh nervous barks to the swelling cacophony of rising fear. And it can result in an explosion of water slapping as they take to the air, incapable of sitting still with such tension bubbling. And if the ducks take flight, they're usually closely followed by the couple of spattering coots, furiously pedalling water until they become airborne. It soon calms though, and the general business of preening and rootling among the bulrushes continues. However, the distressed call of a goose under immediate attack from a predator is unforgettable. It doesn't happen too often, but when it does it rents great gashes through the air. It's a bloodied rasp that halts you mid-breath. It has already happened on two occasions this year, one of which sadly culminated in a bundle of long white feathers appearing at the bottom of the field. Geese are monogamous, and most will stay together for life. If one dies, however, the other will attempt to find a new mate. Sadly, it may have been too late in the season to produce a new brood for these widowers.

With the sky now darkened, the three maples are a hazy stand of woven shadows, barely discernible from the indigo flood beyond. The air has cooled and my hands have found their way to my coat pockets where one finds a special stone, given to me by our youngest child. The pond is quiet. The need for tea and warmth has my feet turning towards home. One more call from the coot strings out across the water.

Yesterday our youngest headed back to school for his new year and with our eldest now at university and our middle child preparing for a year of adventures beyond our waters, I think I know how these feathered parents feel. Pretty exhausted would be the first wave of emotion.

They're growing up and out from this nest, and I'm excited for them all. But I also feel flooded with many conflicting and unexpected emotions. I'm not a very 'cool' mum. I worry a lot, I hug a lot. I've been told – as a comfort perhaps, I'm not sure – that the sense of deep-rooted worry never quite leaves you. Even when they're rearing children of their own. I'm not sure that offers up any kind of solace, but there is a feeling of connectedness and continuity in that thought and that does afford a small but palpable sense of relief.

Field Maple

The Egg

The beginnings of a bittersweet commission: a mistle thrush's egg, heralding a brief but very welcome return to spring. At least in the studio. This year has been in such a hurry, at times almost tripping over itself in its keenness to reach autumn. And now she's here.

My studio sits on the edge of an ancient wood. Among the dips and rises of this stretch of rustling wild land there are huge shoulders and elbows of sandstone that jut out, half-waking giants, shaking off their blankets of leaf mould and moss. My morning commute is a direct path of no more than two minutes. But unless it's raining, this walk will often meander awhile and lead me past the shiny ramble of wild rhododendron to follow one of the deer tracks that slip deep into the woods. This particularly well-worn path weaves between birch and elder, to then steer round and up through a gully banked by two of

Viridian

Oxide of Chromium

Ultramarine

Caput Mortuum

these huge and hulking outcrops. This seam of rock runs for hundreds of miles beneath the gentle curves and dips of southeast England, often cresting the earth to form a ridge like the backbone of some vast and long-forgotten creature. Eventually it gently rolls beneath the sea to cross the Straits of Dover where it rises once more to take in the airs of northern France. A few more strides and the gully opens out onto a hornbeam-flanked glade that mirrors every season's wardrobe, from the first spears of bluebell through to the coppering sea of fallen leaf and bracken.

Outside my studio, a lanky-legged wall of this bracken has once again grown beyond its ability to support its elaborate flounce of green, and so much of it now stands stooped like a gathering of arthritic peacocks, leaning into autumn.

Because of where the studio is perched, the light is forever flexing and shifting depending on the time of day and the seasonal attire of the surrounding trees. As such, it becomes quite exposed and vulnerable to the whims of winter, once she's stripped the leaves and laid bare the surrounding land. I have been known to sit amidst a swaddling of hot-water bottles and blankets, right hand holding a steaming mug of tea, while the left hand busies itself with pen or brush. There's a tiny radiator, but that's more to restrain the damp and creeping fingers of autumn from damaging my stacks of paper.

Throughout the summer months, the two male blackbirds were far too busy (and no doubt exhausted)

Hawthorn

raising brood. But they're back now, like sentries guarding their indistinguishable curtilage of berried elder, hornbeam and hazel.

In the branches above there's a magpie rasping through his limited repertoire of swear words. But as a gentle counter a wood pigeon has begun a soft and mellow correspondence with another, further away among the jaw of pine teeth that edge the forest.

Today I've enjoyed many visitors, both human and otherwise. There've been wasps bumping into the (open) door, drunk and disorderly on the apples that have fallen and begun to ferment at the sturdy feet of their alma mater in the orchard. A southern hawker dragonfly decided to glide in, perform the most perfect square dance, and then glide out to lace among the birch and holly. And then there's the constant chatter of the mice that sensibly decided the felting beneath the roof tiles would make the perfect summer nursery. All, of course, are very welcome as long as they don't chew the paper or, like the occasional fly, decide to leave a crazed dot-dash calling card of poo across my drying work.

I found this little egg along the edge of a large thicket of broom that stands like an impenetrable fortress along the top edge of our field, and we've left it to spread and grow, knowing that a whole raft of birds use it as the perfect haven for nest building and child-rearing. The egg would've been laid any time between February and the first warm breath of summer. In a nest built from grass, roots and moss, and held together with mud, it would've been one of perhaps four or five. Sadly this little egg has a meticulously clean hole on the other side, one likely made by a magpie. But there's a good chance that if the parents started early enough they would've braved a second clutch before the end of the season.

As autumn races ahead, the mistle thrush that laid this egg should be filling her stippled belly with the bountiful supply of slugs and snails, the majority of which seem to have gathered along the edges of my brassica bed, a silent army of slayers awaiting nightfall. There's the last of the adult grasshoppers in the field, and, of course, there's now the chaotic cavort of barely airborne craneflies to tempt any bird looking for an easy snack. Later, as the fauna supply peters out, hawthorn, holly and yew berries will help sustain them through winter months. The avian larder is certainly brimming.

Observing the transition of one season to another can be breathtaking, like revisiting a much-loved dressing-up box, full to bursting with the promise of all that is gaudy and glorious. There's the copper beech that lights up the gloomiest of days, the line of birch that bud and unfurl to a bright green in spring. In late winter, the sprawl of woodland behind my studio is lit up with wood anemone that glow in low light like an earthbound constellation.

But for me, perhaps the most visually splendid and genuinely heart-warming tapestry is the one that unfolds as summer passes the baton to autumn. I'm lucky that from here in my studio I've become the quiet and happy witness to every nuance.

But, of course, underpinning this is the undeniable comfort of the continuity of it all; it has the ability to rein back the stealthy prowl of seasonal melancholy that can grip even the hardiest heart. As day follows night, season follows season.

Holding and ultimately painting this precious egg does much to shore up the promise...

Spring, then summer, will return.

A Handful of Beans

(Written in October 2017, before we took apart the veg garden)

The dawn of a new day, and October has slipped in through the open door, riding on the seemingly endless tide of windswept leaves. The ritual and rhythm of a walk through the field is calling.

There's the added boon of a new pup, Billie, to enjoy this time with me. She's all legs, oversized feet and a mass of brindled fluff, and she reminds me of the joy of just *being* in this sea of green, which can be easily overlooked.

The field is beginning to hunker and huddle for winter's approach, and much rain has tramped down even the sturdiest and most stalwart of grassy outcrops.

The acrobatic grasshoppers have left the stage, but in their place many spiders have been busy weaving huge trampolines between the hollowed husks of the once blowsy petticoats of mallow. Their silken skeins are taut with the anticipation of a meal – the master builders waiting at the edge, one foot poised to feel the slightest ripple of the foolish and fallen.

Cadmium Deep Red

Lake Rose

Yellow Ochre Pale

Raw Umber

Billie is completely oblivious to these small but definitive set changes; her head is down, and her nose is full to brimming with the unruly and chaotic torrent of new scents that clamber for her attention as she snorkels through the dewy grass. There's rabbit (so many), deer and fox (the serial killer that meticulously slaughtered all of our chickens) and also the freshly dug, blackberry-blue of a badger latrine. As a youngster she hasn't quite mastered the skill of self-application, but I can tell that to her it's like raiding the shelves of a perfume counter… Penhaligon's-for-pups. Her eyes are slightly crazed and showing a lot of white.

Onwards to the veg patch. It's not looking its best, but even in this state of semi-decay there are elements that never fail to make my heart happy. The empty hulls of forgotten and spent marrows lie stranded in their earthy bed like washed-up shipwrecks unceremoniously dumped by the careless hand of a storm. The overwintering broccoli are wearing peppered leafy overcoats. Morse-coded with tell-tale dots and dashes… 'the coast is clear stop come dine here stop': evidence of a successful assault by the late army of cabbage whites that I've been batting away. A futile pursuit. But with any luck the temperatures will continue to plummet and so these regular raids will peter out.

To Do
- Remember to remove egg from fridge
- Must order more brushes.

And then there are the glamorous, slow-nodding pompoms of overgrown leeks. With a good friend's encouragement, I decided to leave them and so they're now approaching their third year. I'm dearly hoping for a huge crop of scapes to roast and roll in a thick balsamic vinegar; they go so well with a soft and creamy goat's cheese. But even if this dreamy promise doesn't come to fruition, their evolution has been a visual feast, like watching fireworks in very slow motion.

I make a visit to the greenhouse, to pick some of the late tomatoes that are beginning to glow among the turning leaves. They have an air of Christmas-come-early that makes me smile, but also wince a little; local shops have already started putting out the appropriate cards and walls of family-sized biscuit tins and boxes of chocolates. It's only October, for goodness sake.

Then I spy the drying borlotti pods. Once the plump and dazzling eye candy of the bean bed, they've now become sunken forms; their parched skin no longer fits, like the bagging skin that cloaks the arthritic knuckles of ancient hands. Gathering up a pod, a whispering chatter-rattle escapes from the jostling passengers inside. Without thinking, my thumb happily takes charge and slides down the crisp and gaping hinge point to reveal the perfectly preserved little worlds within.

A few winters ago I sent my mother some dried beans. I didn't know whether she'd actually want to grow them, or indeed if she'd even have room for them in her busy

plot. But during the following spring she sowed them and grew a sprawl of beautiful, splatter-patterned, purple pods. And to my surprise when autumn arrived so did a little brown envelope filled to splitting with a new bundle of borlotti beans.

And so from these I grew my own tangle, and enjoyed some happy, noisy meals with my family.

I duly harvested and sent some of these new beans back down to her. And the following year she grew more plants and did the same.

This year has been wholly productive in the vegetable garden and the borlottis have triumphed once more.

I've saved and dried a few dozen pods. And now, having shelled them all, it's time to send this new generation back down to Mum, for next spring.

This simple relay of sowing, growing, saving and sending, I dearly wish we'd started it way back when... when I realised that growing things was destined to be one of the most pleasurable and enduring pursuits, rather than 'just something your mother does'. It has the ability to buoy me up when I'm flailing and all at sea, and if ever I get a little too smug, it's quick to deliver an earthy slap of reality. The simple and physical hard graft of

growing vegetables – it's perhaps the most honest and direct friendship one can have with the land – is both humbling and inspiring. And for me, the rewards are beyond those that I bring home to the kitchen.

I'm taking these beans back up to the house and will write a note and seal them in a brown envelope.

I know I'm being ridiculous and I'm aware that none of us will ever really have a true grasp of how this world flows, but then this leaves an awful lot of room for a bit of magic. So I'll admit that I have nurtured an undeniably bonkers and perhaps childish hope that if we can keep this going, year in, year out, then karma will smile and make sure that we're both here, each year, to keep the seeds growing.

Winter

Fig

Yesterday we had to cut a branch off our vigorous little fig tree, in order to lever up a thicker, straighter limb and carefully secure her to the south-facing wall of our house.

I loathe removing perfectly healthy branches – or even just pruning the fig tree – so I often leave this job to my husband and walk away feeling quite traitorous. But this side of the house gets a lot of weather. The wind whips around this wall, chasing its tail, picking at sandstone, rubbing edges off the corners. So giving her a helping hand is vital if she's to continue to thrive.

Muttering an apology, I reached deep within her tangle of limbs to hold the branch steady; arms, face and neck became abraded by the last of her huge rasping tongues of yellowed leaves. Along with this branch, there were other, more supple, finer limbs that had shot up from the base, and as we trimmed and tied, the milky sap slowly began to ooze from each cut, and I was transported by the heady scent.

I'm not entirely sure when I became so completely taken with figs; the 'why' is nigh on impossible to explain. Often those who've tried figs fall neatly into lovers and

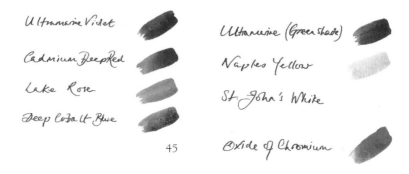

Ultramarine Violet

Cadmium Deep Red

Lake Rose

Deep Cobalt Blue

Ultramarine (Green Shade)

Naples Yellow

St. John's White

Oxide of Chromium

haters, a 'Marmite' moment for many, I suspect. And when confessing my deep-seated affection for these remarkable fruit I often get a look of vague disbelief… which slips into one of mild boredom if I over-prattle.

It's not just the taste. If you're lucky enough to live near a fig tree or perhaps have one growing in your garden (or, indeed, have a blissfully ignorant neighbour whose fig tree has decided to rest an arm casually along your communal boundary) then you'll have had the opportunity to watch its miraculous transformation… from a gnarled, seemingly lifeless grey body, to one of handsome virility.

In spring, from unseen creases in elephantine branches, the tree begins to push out little beaded fruit, at times so small, you may not even notice this little burgeoning army of acid-green blisters. These gradually grow and become stemmed, much like the slow and tentative inflation of a birthday balloon. As summer opens her arms to the promise of balmy weather and longer days, these little fruit continue to swell, peaking out from beneath a fringe of lobed leaves the size and spread of a hand, unfurled and waving.

And then, they ripen.

At this point I could go off grid and wallow thigh deep in The Fig throughout Religious and Cultural History – you know… Adam and Eve, etc. But whether we're genetically pre-programmed to have this connection with the fig tree or not, I 'know' that Eve really should've ignored the serpent and made her way back to that fig tree where she'd hastily gathered up her makeshift knickers. *Surely* there would've been something infinitely more enticing than anything that snake (or indeed Adam!) had to offer. Had she looked a little closer, beyond the curtain of leaves, she might have seen branches laden with heavy, Rubenesque fruit, barely holding on, eager to fall into her outstretched hand. Reading beyond the story we were fed as children, I've found many suggestions that the 'forbidden fruit' could've easily been a fig, grapes or even a lemon. In his glorious depiction of Genesis on the ceiling of the Sistine Chapel, Michelangelo even painted the serpent offering Eve a fig, rather than the infamous deal-breaker apple. And, dare I point out, it also shows Adam being just as keen to pluck one for himself. Understandably so.

How to eat a fig: no one explains it quite as well (and with a glorious disregard for blushes) as D H Lawrence:

> *The proper way to eat a fig, in society,*
> *Is to split it in four, holding it by the stump,*
> *And open it, so that it is a glittering, rosy, moist, honied,*
> *heavy-petalled four-petalled flower.*
> *Then you throw away the skin*
> *Which is just like a four-sepalled calyx,*
> *After you have taken off the blossom with your lips.*
>
> *But the vulgar way*
> *Is just to put your mouth to the crack, and take out the*
> *flesh in one bite.*

The flavour and texture of a fig, when picked ripe and wearing a scarlet smile is perhaps the most joyful and shocking food experience I've ever had the pleasure of discovering for myself. It is at once both innocently sweet and unapologetically sensuous. And, of course, you can peel it, cut it up, render it entirely unrecognisable, but a fig will always be a fig.

Perhaps the best I can do is capture it on cartridge, using egg tempera. This is a lengthy, involved and age-old medium that requires an egg yolk, ground coloured powders, and a lot of patience. But the resultant artwork will have a lustre and depth that I find utterly beguiling. So it would seem a wholly appropriate medium to use when portraying such a heavenly body.

Meanwhile, I'm standing outside with the full blast of winter now clawing at my coat and scarf-muffled face. It has become eye-wateringly bitter today but I need to clear away the cut branch and tangle of shoots lying at the base of my fig tree. Sadly, I can see there are many hard, green, miniature grenades that will never ripen. But I've found a recipe to preserve them whole in a spiced syrup, and this will give much joy during these colder months. Gathering up the scattering of grey limbs, I carry them to the wheelbarrow only to become engulfed in their perfume once more. Placing the branches in the barrow, I take out my secateurs and cut a scented stub to keep in my coat pocket. It'll stay there until next year.

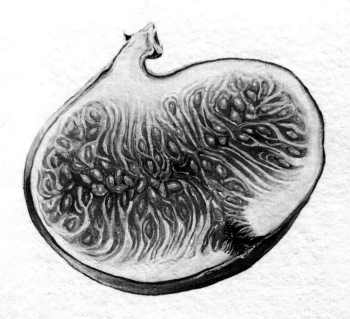

Naked

Today I woke to the brightest of winter mornings, the kind
that makes your eyes water when you venture out, not
because it's windy, nor cold, as it is both of these things,
but because of the colour of the sky. The day is shorter of
course, but the quality of each hour, indeed the minutes of
a day such as this are almost enough to feed a winter-weary
heart for the next inevitable bout of faceless grey.

Stepping outside, layered up like the Michelin Man,
I instinctively began running through my various
squeezed tubes of oil paint names for different hues of
blue, desperate to classify this sky. And, of course, every
paint maker – Sennelier, Winsor & Newton et al – has its
own version of each colour. No two Ultramarines are the
same! But I think if I'd had to name it, it would've been

phthalo blue

cerulean blue

ultramarine blue

cadmium blue ;
Saint John's White

51

Rembrandt's oil, in Phthalo Blue Red. This winter sky was the kind of blue that feels wholly unnatural, the kind that throughout history men have become maddened by and have spent lifetimes and indeed lost fortunes in their efforts to recreate.

Blue was once considered one of the most revered and elusive colours. Egyptian alchemists played with the semi-precious stone lapis lazuli to create blue pigments that only royalty could afford. Roman Catholics commandeered it for colour-coding their saints, and the Virgin Mary was given robes of blue to symbolise her innocence and trustworthiness. More recently the artist Yves Klein pursued the creation of a unique and singular blue. He even painted the bodies of people in his shade, International Klein Blue. And, of course, Miles Davis allowed us to dream awhile enveloped in his own sweet evocation of this colour. I think the richest blues still have the power to evoke an involuntary response in us all, such is the beguiling and almost unobtainable purity of this colour.

And yet, there I was, strolling in the frost-dipped field towards our woods, enjoying this extraordinary sky for free.

Winter can tend to be a busy time in our woods. When the ground is hard and all growth is in hibernation, it's the perfect time to head in and start cutting, logging and hauling future firewood. Into spring and we have to worry about bluebells, ferns and generally disturbing breeding wildlife. Later on, when the bracken and bramble have filled almost every dip and hollow, everywhere is a riot of growth and, other than the usual paths, most of the wood becomes inaccessible to us humans.

It was quiet today, though, and I had the wood to myself.

For many, wooded areas can feel wholly unapproachable, sunless places full of an unfathomable potential to disorientate, scare perhaps, and overwhelm even the most confident among us. Every adult can remember a bedtime story in which a child gets lost/eaten/kidnapped/turned into something ungodly to haunt the next unwise wanderer. And they nearly always involve a wood. A deep, dark wood. But, in truth, they're just places where trees have chosen to take root and make their home.

When we first moved here, we had just two very young children and I dreamed of being able to play hide and seek with them, climb trees, make dens, spy on wildlife, and have many picnics, of course. But the truth is that it was a pretty inhospitable place, and they didn't much like to venture further than the first guard oak. It took us a fair few years to open it out and encourage light and wildlife back, and, during that time, our two had become three, and a little older, and a bit more confident. As they grew braver, the woods became enchanting, irresistible. It fired their imagination and became the ultimate playground for their adventures.

We still have the remnants of log circles where little fires were built and marshmallows were melted. There's a small tarpaulin that used to hang low between two rhododendron branches poking out from a sandstone nook; a place to read books, make stick weapons and eat chocolate biscuits. There are the remains of some old china tiles and a reclaimed birdbath that my two eldest transformed into a fairy garden. And everything will stay exactly where it is.

At this time of year the trees stand leafless; every outstretched arm is laid bare, their secret clutches of old nests exposed. Birds perch, puffed and mostly silent until the sky begins to lift. And with all the bracken having rusted and fallen in on itself, the wood lies open with her venal spread of paths and tracks now clearly visible. And so I've learnt the best time to explore any wood is when they're at their most empty, when the low light can permeate even the deepest corners, and it's now, in midwinter.

Leaving the dogs indoors meant I could be more considered with my footfall, and hopefully a lot quieter. Trying hard to avoid snagging my husband's coat on low branches, I made my way between some self-seeded young hazel whips and took the deer-rutted path that leads to our pond. There were noticeable patches of deeply grooved edges where the deer pause and brace themselves to sip at the water. Looking into the calm mirror of today's sky, I spotted slow, shifting clouds of frogspawn that reminded me of one of my most favourite school puddings, tapioca (always with a spoonful of strawberry jam. Don't judge me).

Moving on past a stand of birch and looking down among the compressed carpet of fallen leaves and rotting branches I spied the small wizened shoulders of the ferns. It'll be quite a few weeks before we begin to see the new young fiddleheads appear but it's a comfort to know they continue to thrive in spite of the marauding army of bracken that, come summer, threatens to swamp everything in its path.

Through years of leaf-fall, decomposing trees, the gradual desiccation of the sandstone vein that runs throughout, and the many small springs that continue to bubble, the soil here is rich, dark and fertile. And where I was standing, would soon become a tightly luxurious sweep of green... the beginnings of the bluebell carpet that would unfurl, come late spring.

No matter the season, the rhododendrons are an ever-present canopy of green and offer some relief to an otherwise rusted stage. But it does seem that every year they work in unison with the bracken, encroaching just a little further into carefully cleared pockets of quiet that we've tried to recreate for wildlife to thrive. I think, given the chance, together they would happily throttle this woodland and swamp all the light and life that now seems to flourish here.

Checking the time, I took a smaller, curving spur off the main deer track, that switched back in the direction of my studio and I found a casualty of the strong northwesterlies that barged through a couple of days ago. With carefully woven twigs, dead leaves, much knitted moss and a little bracken, it comfortably filled my open hand... a wren's nest. And clearly the builder of this small marvel (the cock bird) had struck lucky as this one was lined with feathers. He may have built quite a few, but the hen

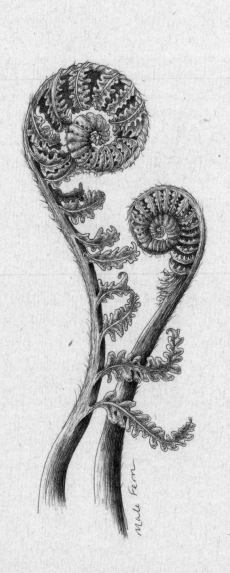

Male Fern

bird had decided upon this one as the winner, and had accordingly lined the small, deep cup with feathers. So this was a nest that had held and possibly hatched a new little generation. Hopefully they fared well last year and will be back again, to breed come April.

Heading upwards and into the open glade at the edge of my studio, I turned and looked back down, tracing the various paths that had brought me here; a network of journeying, the to and fro of the hooved and clawed, all of it gilded by this morning's sun, many lifelines lit. I've tried very hard not to make too many tracks of my own through these woods. Inevitably we've had to, but those paths we've carved out I hope will transpire to do more good than harm in the evolution of this bit of wild.

Crow

It's late morning in early winter, and I've decided to take a break from the studio to give the dogs a run around, and my eyes a different length of focus. Working at close range on a detailed illustration can be quite tiring, and stepping out of the door to look beyond my window to the edgelands of our wood is both a salve and a barely hidden joy. Grabbing a rather tired, de-fuzzed tennis ball, we make our way down the path towards my husband Marc's log store. A blackbird explodes from the bank of skeletal broom and a couple of rabbits dive into the mesh of bramble opposite to lie still until we've passed. There's a fresh pile of woodchip from a recent bout of sawing logs, ready for splitting, and it smells slightly fruity – vinegary almost. I think it's from some fallen birch that needed clearing. The scent hits the back of my nose and transports me to a particular day when I must've been no more than nine years old.

I'm back in Cornwall, deep in the forest with my father. He's wearing a brushed-cotton shirt, sleeves rolled up; the muted check spreads comfortably across his back as he's bent over, chainsaw in hand, working on the trunk of an old disease-ridden elm. I'm waiting behind him, where I've been told to stand, watching the flecks of fresh sawdust arcing up in a fine, sweet-smelling cloud to settle around his scuffed and scarred work boots like confetti. The dark, root-run floor has been dressed in a fresh, new cream-coloured carpet.

It's mid-winter, and the predominantly pine forest is lit by the low-slung sun that's managed to slip through among the tall and ancient trunks. It feels bright, busy and exhilarating to be out here, stamping feet and clapping hands, helping my dad. I'm feeling particularly chuffed because I've noticed that today he's wearing his army green, 100 per cent nylon 'indestructible socks' that I bought from an advert in the *Sunday Times*. The ultimate socks: 'Genuinely indestructible, or your money back'. Surely the BEST present, ever!

The chainsaw splutters and all at once it's deafeningly silent. Dad looks down, brushes something off the leg of his jeans, then with one jerk starts up the chainsaw again, legs braced and standing sturdy as he slips the teeth of the chainsaw into the tree trunk once more. I think nothing of it and it doesn't occur to me to ask what just happened, until I come around to stand to the side as we begin to grab at the rough damp chunks of wood and load up a wheelbarrow. There's a crimson creep of blood seeping through a freshly torn gape in his jeans. Well he's not bothered, so I shouldn't be either. I happily ferry the barrow over to the back of the van that's parked up on the old lane and begin to hurl the sweet-smelling logs in as far as I can, enjoying the heavy thud as each one lands and skids across the rusted floor to clunk the foot of the driver's seat.

We carry on this way, cutting, pausing and piling for another half an hour or so. We're both getting hot. Dad's face has begun to shine with the exertion, but I've noticed his ruddied cheeks are fading as a white flush begins to spread downwards from his brow. It's not a look I'm familiar with and I later learnt from my mum that it was the delayed shock and 'what-ifs' that had finally taken hold.

But never one to fuss unless it's serious, he maintained, when I asked him about this recently, he was simply annoyed that he'd ruined his only decent pair of Levi's.

'It's time to head back up', he smiles, not once looking down to inspect the wound that's still spreading a bloom of red across his jeans.

Overhead a single crow lets out a rasped and throaty 'caw'. To my young ear he sounds reproachful, almost mocking. We heave the wheelbarrow into the back of the van, and tie the doors shut with a length of binder twine. Jumping up and shuffling along onto the front seat, I give Dad a beam. He's covered in sawdust – it has lodged in his hair and eyebrows. Father Christmas come early! Edging our way back up the lane, Dad chatters about mugs of tea and Mum's fresh lemon drizzle cake, and my hands are warming at the thought. The jeering crow is forgotten.

My father was the complete opposite of me, aged 9 but going on 99. Everything used to worry me. A tiny paper cut would have me howling, racing to the kitchen table to await love, sympathy and a plaster. A 'firm' word

Indigo

Manganese Violet

Cerulean Blue

Ultramarine

Mars Black

Oxide of Chromium

from either parent would result in a sense of unworthy doom descending on my young head. Any joke aimed at me would feel as sharp and pointed as blackthorn. And a comedy show on the telly might have them holding bellies with tears streaming, while I would sit there looking from face to face, trying to fish out the joke from the deep unfathomable waters of *Dad's Army* or *The Good Life*. I struggled.

I'm not quite sure when I discovered the 'lighter' side to life, but I suspect it leaked in through the cracks around the same time that I hit puberty, grew boobs and had my first kiss. (Make that the second kiss. The first was a disaster!) Things definitely improved and my outlook became sunnier!

Our children have each taken their own unique path through childhood and into puberty, and I marvel at the apparent ease with which they edge, straddle and jump most of the pitfalls that seem to beleaguer us all in these formative years. Growing pains – nothing to do with the aching bones as they race the skin to full height, but everything to do with navigating one's route to adulthood and beyond. As a parent now, I can imagine mine had a tough time trying to steer me through this maelstrom.

But they did, and with much love and a seemingly endless reserve of patience.

We've weaved our way among the bare trunks of quiet trees and I've spotted two deer. Although we do get muntjac here and very occasionally roe, the predominant sightings in our area of Sussex are mostly of fallow deer. These two are female, looking lighter and a little more fragile than the bucks. In summer, their colouring is the warmest butterscotch with a light dusting of snow. Today they're wearing their winter coat of deepest taupe. They look young, probably last year's fauns. It takes them all of two heart beats to spot me, and bolt.

Oblivious to this moment, the dogs have disappeared into a thicket of leaning bracken in search of other deer that occasionally make this their bed. And although Billie, a lurcher, has more of the frame for an impromptu game of chase, the two dogs are never in luck and from a crouched start they don't stand a chance.

It's time to head back to the studio.

The winds have picked up and a quiet whine and creak of conversation has begun among a stand of silver birch nearby. Above me now, there's the resident crow who prefers to sit atop a swaying and rather sickly ash that edges the field. He lets rip a cry into the grey acre of sky. And to my older, happier ear he sounds like a cheerful scaffolder possibly appreciating the view below.

Morning Post

Oh February, first you walk mud through the house, leaving a clumped and wet trail of winter's tilth. Then you tease us with small and beloved envoys of spring.

The crocuses are beginning to stretch their pale, swan necks, wide-open beaks showing yolk-yellow throats, and young, sunny-faced primroses can be found peeking tentatively from among the teeth of last year's bramble, only to become wind-whipped and drowned beneath a deluge of yet more rain.

And my god, what rain.

My bees are utterly confused with the sudden early call to arms. With their queens still in slumber, recovering from the previous year's marathon, they're unprepared and there are signs of mild panic. Occasional clouds of pollen can be seen lifting on the wind like a fine curtain of mustard powder from the early-formed tails of hazel and alder catkins. But it's only the brave and foolhardy bee that leaves the warmth of the hive and tries her luck during the wet and wild days that have chased us into this month.

In my efforts to expend some of our lurcher Billie's endless supply of bounce before bedtime, I've taken to grabbing a torch and walking down through the field at night. This evening the ground is literally bubbling with

Crocus

each footfall, the earth incapable of swallowing any more of this endless rain. There's no moon visible to backlight the fresh herd of heavy clouds that have gathered as if to plan tomorrow's deluge. But the air smells clean. It's good to be out. Rather predictably, my feet steer me on my usual route, through the tired and bumpy islands of tangled, rain-flattened grasses, to the gate of the vegetable garden.

Reaching out for the latch, I remember there *is* no latch – there's actually no gate, or fencing.

We built our vegetable garden some 15 years ago and it reached a point where there was more wildlife within the enclosure than outside. Each visit brought fresh waves of frustration as I'd find the evidence of a cloven-hooved rave or the remains of a midnight feast – the young sweetcorn and pea pods proving to be the most popular bar snacks. The fencing was rotting, its netting dug up or pushed aside by rabbits, deer and, more recently, badgers. The partition boards had eventually succumbed to old age, and the soil had almost shrunk to half the original quantity – hardly beds any longer, more threadbare carpets. The only things that had stood firm over time were the brick paths that we laid with our children. But as gorgeous as they looked, I'd at last had to concede that they were far from practical – full on weeds in the warmer months (strimming them only, to avoid weed killers) and a skating rink with the arrival of first frosts in winter.

It had needed a rethink. Last year we began to dismantle the entire plot and suddenly I had nowhere to grow our vegetables. So I didn't order seeds. The impact on my sense of purpose was a shock. I can't quite believe how bereft and ungrounded it made me feel. Planning, sowing, planting out and nurturing our crops gave a structure and rhythm to my year that I'd barely noticed, and so I found it entirely unfamiliar, even hard at times. I lost my drive. Most of what I illustrate comes directly from the soil. Much of my cooking is inspired, almost dictated, by whatever I've unearthed among the veg beds. I really hadn't imagined that the person I'd become was so inherently entwined with the growing of our food.

Tucking my left hand back into my coat pocket, I raise my right hand and slide the torchlight over the vast expanse of mud that will become my new beds, and my hearts sinks at the amount of work to be done before I can plant again. But the beam stretches to the far end and finds the bare bones of huddled pea canes leaning against the hedge, and it's just enough to raise hope, and the corners of my mouth. It's been the soggiest of Januarys and everywhere is depressingly waterlogged. But perhaps this bodes well for a lusty surge of verdurous growth, come spring.

I turn and call Billie. She emerges from a pocket of night, a slightly less crazed look about her, and we're heading back home now, through the wet field, when the torch light snags on the early embryonic leaves of field mallow, and then the tiniest lobes of the borage that we sowed last year.

Primrose

And with all the enthusiasm of the blackbird that strikes up every morning as the light teases at the far edges of the forest, I find my bounce. Billie thinks I've gone nuts but is more than happy to play along, as apparently bounce is a lot more fun than trudge. And to be honest, this winter there's been a fair bit of trudge.

I've friends who relish everything about this season, and I, too, adore the time and space it allows for quiet and contemplation, with perhaps a measure of single malt to warm one's thoughts. With its darker, colder hours we're freed up from the frenetic level of activity that we've come to associate with spring, then summer. There's the comforting whiff of a stew or a roast that can slip up the stairs and lurk for days. And with the scarcity of daylight there's a unique and exquisite touch that this season brings, when the sky cracks to let a finger of sun slip through its vast acreage of grey; the way it gilds all that it touches is unsurpassed in its beauty. Nevertheless, I think even though we may relish these delights, most of us do reach a point when the sense of drowning in the heavily knitted cocoon of winter becomes too much.

There's a collective sigh when we note the incremental lengthening of each day, and shoulders relax and necks emerge with the steady creep of mercury. The first froth of roadside blackthorn waves us onwards, and the heady perfume of daffodils brings much cheer. But for those that grow things, the shift towards the new season is signalled by the arrival of the seed packets.

Billie and I are now back home, and while I wait for the kettle to boil, I'm thumbing through this morning's post, which has been left until now. Among junk mail and the brown envelopes with text glaring through plastic windows there's a light, padded envelope that – on squeezing – feels irregular. The kettle's forgotten as it begins to mither because I'm ripping open the package. A dozen seed packets spill out onto the kitchen counter... several kinds of tomato, pepper, chilli, pea, broad bean, courgette, corn, radish and finally a favourite to eat and illustrate: beetroot. I've missed beetroot a lot. Instinctively, I scoop them up and hold them to my chest.

These small, rattling packets feel like tickets to summer.

Rebuilding this vegetable plot is beyond daunting, but I am excited to be starting again, and with any luck the decisions we make will be a little wiser after our 15 years of sowing and growing.

Spring

Finding Spring

We're galloping through March but, by contrast, the creep
towards spring equinox has felt painfully slow. With
continuous stretches of relentless rain and shoving gales,
it always amazes me that the blackbird exhibits such a
remarkable persistence to make his voice heard each and
every morning. Taking his lead, I head out at first light
most days to seek out evidence of the steady rise of sap, as
spring begins to spread across the field and wood. I make
mental notes and often jab away on my phone, detailing
observations. It can often feel a bit like signposting my way
to each season and it gives much comfort when progress
feels thwarted.

Blackbird ~
Turdus merula

loves worms, insects,
berries, also apples from
the orchard!

Yesterday, while shuffling through some papers in my studio, I came across a diary entry from two years ago, when spring became frozen mid-step, and the world became hushed:

10th March

> *It began with enough snow to wholly cloak our little corner of the world in white. Once this crystalline blanket had settled, all noise became muffled, with only the most prominent features able to rise above the swaddling of snow. And we became housebound for three days.*
>
> *On the first day I made a slow and squeaky hike through the field to the veg beds, to find the plot dressed as some huge, badly made bed, its sheet barely tucked at the edges, with more than a few old marrow bodies slumbering beneath. I looked beyond and noticed, for the first time, the stripped-back woody stems of forgotten purple cabbages; reaching out from their sea of snow, they appeared as the tentacles of some exotic red-armed octopus, buried but still flailing.*
>
> *On to the tiny orchard to check on the new, young plum and pear trees, only to find their little arms bravely basketing at least their bodyweight in snow. A gentle shake and they were free to wave once more.*

I realise that birds don't have the luxury of marvelling at the 'beauty of it all', and no more clearly was this proven out than watching two male blackbirds, perched and punchy, oblivious to the maelstrom of winter in flux, and slinging verbal abuse at one another. It may have had something to do with the last remaining baubles hanging from the crab-apple tree.

With the dogs almost in tow (somewhere, but most likely rootling at a freshly laid cluster of deer poo), I carved a path across the bottom of the field. Through the fringe of trees, I spied a gathering of Canada geese and mallard, two lofty and nervous cygnet swans edging the group. All of their usual moorings in the lake were partially frozen, so territorial lines had to be renegotiated, and possibly crossed, in order to find food and new anchorage.

Through the woods we creaked, weaving among birch and beech, my one footfall for every two of deer, branding deep into untouched powder. A startled cloud of puffed-up pigeons lifted as one from their lanky ash, clearly not expecting visitors in this weather.

must stock up on ~
oxide of chromium
sap green
Naples yellow light

This is my daily ritual throughout the year,
and sharing it with an old terrier (Chewie) and
a lurcher (Billie) is wonderful – if a bit of a pain
when I'm hoping to listen out, and spot any new
feathered or furred guests in the woods. However,
during the white-out days… watching Billie
in raptures, as she discovered the Marvellous
Circus of Snow that had filled every bluff and
badger hole, was a complete joy. I found myself
laughing madly, and joining in. Which, of course,
seriously confounded the terrier. Humans just
don't do that unless they're small and loud with
grabby hands. And those kinds of humans are best
avoided at all times.

Forging onwards, we made it to my little studio.
Negotiating the slippery wooden deck is always a
bit of a challenge. I'd carefully placed a scavenged
beach rock on the one sagging board so I wouldn't
step on it and bring it to an untimely end. But Billie
likes stones and had picked it up and taken it to her
place-where-all-special-things-must-go. We made
it in, board intact, slid the wedge of fallen snow
back out, and shut the door. Settling onto chairs,
cushions and into corners, we all got on with our
usual routine… drawing for me, sleep for them.

Watching a season's progress through my
studio window is inspiring, and admittedly quite
distracting. Every nuance of colour as it blooms

*and fades has me running through pigment
names in my head, making up new ones. I'd like
to fix a camera up and record a whole year of
this kaleidoscope.*

*During these snowy days this little studio flared
to an almost otherworldly state of bright. The
hours of work played out till way beyond the
usual sundown, such was the lingering gaze of
reflected luminosity. But by the second day on this
'island', the weather turned. A gathering grey of
raging winds stropped in from the east, grabbed
great handfuls of snow and began hurling them to
fly like torn bedlinen across the glade beyond my
studio. By sunset a spill of indanthrene blue had
flooded the sky to brimming. After turning off the
studio lights, we made our way down through the
snowy clumps of rusted bracken to the edge of the
field, finding it stirred and blurred by the gusting
gales, yet it still retained its lucent glow like some
deep-dwelling creature of the sea. We could've
been anywhere.*

*I'm writing this having spent a very early
morning in dressing gown and wellies, walking
with the dogs across the field and into the woods.*

*Everything has become green again and it
appears that, rather than killing off any new
growth, all infant borage, mallow and knapweed
seem to have survived and thrived beneath their*

temporary blanket of snow. Beyond the birch
line, the geese, ducks and swans are once more
spread wide across the dark, still canvas of water,
no longer forced into close proximity by a prison
of icy confinement. There's a level of noise that
is almost deafening, such is their enthusiasm for
open debate.

All is vivid, visible, and almost shouting out for
spring to hurry now.

And, of course, I feel the same. I'm more than
ready for the cacophony and clamour of chatter
and colour that she will bring.

But what a grand finale that was.

Looking back, it's true we were stranded, but I'd
forgotten what peace of mind this enforced aspect of
'islanding' can bring.

A Guardian of Bees

On a balmy Sunday in April 2014 a sweet lady called Pat arrived with 15,000 bees and their queen.

We all dressed up in varying degrees of protected readiness and made our way down to where we'd set up a hive for our new 'friends' to call home. I readied the smoker, nearly burning my fingers as I lit the newspaper, and gradually stuffed the inner cylinder with the dried grass that Pat had collected on our way down through the field. With a suck and puff of air pulled through the bellows, the flicker-flame, slowed and steadied by the compacted grass, continued to burn at a simmer. And eventually, with a little more encouragement, a thick creamy coil of smoke curled lazily from the spout.

Ready.

Pat opened the lid of their travel box and with gentle, slow and loving attention to every nuance of bee behaviour, she removed the six waxed and drawn frames. They were loaded with bees, drawn cells, uncovered babies and capped worker and drone cells. A few cells were filled with honey – their supply until they brought in fresh nectar from their new surroundings. We gradually transferred all of the frames in exactly the same order, pulled cells on one frame rippling intimately into the perfect curve of its companion frame.

The children stepped back a little.

I couldn't stop looking. Shocked, entranced, deliriously happy, humbled. I wanted to take off my gloves (ridiculous) and stroke frames, cells, bees (really ridiculous since although they appeared calm, they were preoccupied and a little agitated by their three-hour journey from Gloucestershire). The smell of the pollen, wax, cedar frames, propolis – the headiest of cocktails.

We stood watching as a few courageous bees began to exit the hive, backwards, hover a couple of inches, land, and then stick their little bottoms in the air, almost performing a come-hither waggle! I learned that this is their way of encouraging their comrades to return: 'This is where your queen is. Stay close, this is home!'

A steady stream of 'wows' began popping from our mouths as we stood, limp-armed and staring. But it was time to let them settle in and get on with the job of locating their nearest source of food. We gently brushed one another down, removing any little interrogative foragers from our suits to ensure none got lost, and made our way back up the field.

Looking back, I can still feel that sheer bubble of excitement. I'd been wanting bees to be part of our lives for a long while, and that day I could barely contain my enthusiasm. If it hadn't been for dear, wise Pat telling me to keep calm and just 'chat to them', I suspect things may have gone very differently.

I'm standing by my hives now and it is another of those perfect spring mornings.

I have two hives now. The occupants inside can be heard
even from a short distance away. It's something I didn't
really know I'd picked up, but I can now tell the mood
of the colony from the hum. I can hear when they're just
getting started for the day, or if they're simply busy and
quite content, or even if they're distressed. I can also hear
the change of pitch that indicates they're making plans.
And around this time, and right through till late August,
those plans can include nurturing a new queen to allow the
colony to divide, with the mature queen flying off with half
of her subjects to find a new home.

This morning though, they sound content, and the
steady flow of bees, to and fro, back-leg baskets stuffed

with pollen, is a happy sight. It means that the queen is laying, that there are young to care for, and that there's food for them to do this with.

I wish them good morning, out loud, and tell them I've come to check they're okay. I pause and then very carefully lever the lid from the hive.

That scent. If I had to choose it would be in my top five, along with the smell of my children's skin, crushed tomato leaves, the space between Billie's ears and the scent of the sea as it pulls at the pebbles. Oh, and the sweet, dusty yet metallic scent of a cut fig branch. That's six, I know.

Every time I visit, I make notes of what I've found. It acts as a guide to prompt me if the colony needs a little help.

They also include details of what's in bloom, and I've come to learn the colours of the pollen that link to those flowers.

Today's notes:

Crab apple
Pear
Apple
Plum
Blackthorn
Cherry
Alder

Visible, audible flow of bees between blossom and hive.
Must number frames in brood box…

Observations per frame

1. Workers cells o/c (open/closed), honey.
   One queen cell open, empty
   Forager dance, captivating!

2. Workers cells o/c

3. Ditto
   Queen, first sighting!
   Large, voluptuous, polished… like vintage Airstream…
   alternately trampled and caressed by attentive harem

4. Worker and drone cells o/c

5. One queen cell; a rice-grain egg, in limbo

6. Honey and worker cells o/c

7. Honey filled, heavy

8. Rainbow stores of pollen, stacking up like a game of 'four in a row' gone mad

9. Empty

10. Empty
Smoker gone out, bees remain calm, very inquisitive.

As am I. Throughout the inspection we continue to chat. Well, they hum and I talk. Often I tell them what I'm doing, what I've found; I congratulate them on their beautiful nursery filled with young, and marvel at the scent of their larder that's filling so quickly with nectar. The conversations have been known to include discussions on child-rearing, both theirs and mine, worries about work. And then there are some topics that are rarely aired to anyone else.

Still chatting, I gently replace the lid, mindful of any unnecessary jolt or knock that would send them spilling out of the hive to see what I'm up to.

It's now almost five years on from that surreal first day, and I've grown up, a lot.

I've changed my approach to growing vegetables, planting more that produce flowers, such as broad beans, peas, courgettes. I leave brassicas to bolt and produce great clouds of yellow blooms. I put down seed trays of water along the paths between beds to enable the bees to drink, in between their foraging trips, particularly on hot summer days. I now sow strips of flowers that make the bees happy, which makes me happy too. I've binned all forms of insecticides, herbicides and pesticides and could probably write the A to Z of organic slug management (still not entirely winning!). I've watched farmers sow vast crops of oilseed rape and then deliberately decimate the hedgerow boundaries to make bigger fields for more commercially geared production and harvesting. A quick fix for the bees, but ruinous for diversity of wildflower and so wildlife. We all know the story of what will happen if the bees are gone. It's a bitter truth, but it represents the mere tip of the sting. Everything we do, and don't do, has an impact on our environment, often invisible until years later. Bees are just one aspect of this story, and they're just one type of pollinator (there are many better), but they're a fair measure of how well we're all doing – or not.

It's not my place to preach, and I shan't. But my eyes have been opened. All I can do is hope to do the best I can by the bees (and all other pollinators – even the loathsome horsefly). Be their guardian, not their keeper.

The Company of Wolves
(Or at least their descendants)

Up early, and out with the dogs. They've both taken their favourite line down to the bottom of the field, carving a dark streak through the dewy grass. They're off in search of beasts both real and spectral. I'm hugging a mug of tea, and watching the new day sift through the last vestiges of night.

It's hotly debated as to just when, on our evolutionary journey, we decided to make dogs part of our lives. It's even less clear when dogs decided that it was a workable plan. Whenever it was (some say it's as far back as 20,000 years ago) it's almost impossible to imagine a time when the relationship didn't exist in some form.

In the beginning, the co-dependency probably relied on the understanding that both parties needed to kill to eat (veganism wasn't necessarily an option then) and perhaps it would work more effectively if they operated as a team. As long as the dog helped, then there was a strong chance that it would get a share of the kill. So food was perhaps the trigger point for this enduring relationship.

If I look at our two dogs now, and if I am completely honest with myself, the truth is perhaps that little has changed. We offer them food and shelter, and they have evolved over time to be less and less inclined to search for their own supply… with the exception of some breeds

Jackdaw –
Corvus monedula
Loves to eat everything,
including apples from
the orchard.
Seems like everybody loves
the apples!

91

that will still happily hunt for living meals, on hoof, paw and wing. Chewie, our old terrier, would love to be quick enough to dine out on rabbit but as yet hasn't quite managed to tick that box. And Billie, the lurcher, would like nothing better than to keep pace with any fluffy, moving target. Also crows – she has a thing about crows.

But, over time, I've come to understand that the relationship I have with these two goes way beyond the food bowl, the hurling of a ball, and the gratifying nose nudge when a tickle behind the ears is required.

Because of where we live, we're surrounded by countryside that reflects each flow and shift through the months in a visual feast of colours and textures. I walk daily and am happy to spy and note down any new findings as the set changes and a new season evolves.

But with a dog (or two) by my side I've come to appreciate a new – and previously obscured – layer of life happening. The human eye, even when open wide, misses much.

Weaving my way down through their narrow blaze of young bent grasses, I follow the dogs to their first pause, close to the corner of the bottom gate that opens inwards from the old lane. It's a fresh badger latrine. Badgers are incredibly tidy animals that dig shallow pits in which to poo; these are often used to give a clear message to other badgers that this is their territorial boundary... So move along. I moved along, with a small bubble of hope rising. Love badgers.

Onwards now. Once more picking up the trail of
darkened grasses, I follow the dogs' tracks along the
millpond's boundary that's lined with a leaning audience
of oak and birch. A small gathering of Canada geese give
a start, and slowly paddle away from their cove towards a
clear patch of lit water.

I'm almost at the east corner of the field where there's
a huge carcass of an ancient beech, and I've caught up
with the dogs. Tail up and marking time like some crazed
metronome, Chewie is digging frantically beneath one of
the elephantine limbs; Billie is standing close by, unsure
of what the plan is. Beyond the patter of earth fall from
each new arc of soil, I can hear nothing. But I can see
rabbit droppings. I call them both to move on. Chewie
reluctantly concedes defeat, for now, and I follow them into
the woods, stooping beneath a branch of hazel that releases
a light dusting of pollen from its lamb's tail catkins. The
dogs' noses have been snagged by another scent and so
they're off among the tangle of tired brambles that are now
flecked with wood anemone. Turning back, I catch sight
of the rabbit that had been hiding all the while, among
the many arms of the old tree. It has a change of heart and
shoots back in beneath the knit of shadows. Probably wise.

Onwards, to where a spring trails a silvered line through
a bed of silted sandstone and fallen leaves, and the ground
begins to pull at each footfall. This steady trickle heads
to the bulrush beds that each year creep a little further
into the old millpond. The dogs have gathered around

the broken, almost lucent remains of a goose egg. There's no nest in sight, and it's unlikely they would've been attracted to the egg unless it had been previously broken. It's probable that the egg would've been pinched by a fox who carried it here to scoff its protein-rich contents in this secluded area of shade.

Turning back towards home, following the dogs past the green mat of young bluebell leaves, and Billie stops, her face set, tail rigid, eyes utterly focused on a collapsed thicket of bracken. I would've walked straight past. But she has seen something. There's an almighty explosion… and fur, hoof and hound are moving as one. Three roe deer are almost gliding across the open glade. There's a dried riverbed and they seem to float over, with Billie in pursuit, and within seconds they're invisible among the trees. It's pointless calling her back, as I've learnt she becomes selectively deaf when it comes to roe deer. A few minutes later Billie is back, cloaked from nose to tail in ancient mud, wearing an expression of exhaustion and deep joy.

Home now and I call Dad to check on the colouring of goose eggs – I think Canada geese lay eggs that are a slightly dirtier white than those of the domestic goose.

My dear father. Some might know him to be a genius at the detailed and exacting task of antique furniture restoration. Those who know him a little better will also know that he is a mender of broken birds. If ever I've a bird-related question, then my first reference is usually him, not Google.

It's a Canada goose's egg. And while we continue to chatter about his own gaggle of domestic geese, a pigeon apparently rises from its nest of yesterday's *Daily Telegraph* and struts along the kitchen table to stand in front of my father, demanding more corn. There's also a jackdaw perched on the stable door that leads to the garden. The jackdaw is wild but has seemingly grown rather fond of Dad. He's turned out to be a kind and considerate landlord, building at least a dozen nesting holes in the eaves of their old millhouse, and a further handful in the roofline of Mum's studio. Under the table is their terrier, who has taken this all in his small stride, rather than his jaws, and actually sleeps with the pigeon alongside the Rayburn, while she recovers from a harsh beating dished out by the more thuggish members of her family.

Dog washed (and muddied clothes changed) and we're heading back out and up to the studio. Billie's at my side, nose gently nudging at my left hand, almost apologetic, though probably not. It's time to get to work; I've an egg to draw, and the dogs have new scents to track.

And there's always a bowlful of dog biscuits in case they get hungry.

Brock Lane

Spring has most definitely sprung. The hives are humming and belching out big plumes of happy, hungry bees. The field maples are unfurling, with great fans of green emerging. The blackthorn has been joined by wild cherry and small tentative puffs of early hawthorn. Frivolous floral popsicles of lady's smock are gathering along some of my favourite roadside banks, and 'my' secret verge of ramsons is starting to overflow with such a generous bristle of spears that I'm almost tempted to tell all of its location... almost.

Another sign of her arrival is that we seem to spend most of each day changing in and out of wet clothes and sodden boots, often clutching bundles of dripping washing, hastily unpegged from the line as a fresh deluge of rain is released from the grumble of sky above. As a season of supposed hope and joy, spring is wearing this weather like a thrift-shop find of pre-loved trousers, and she didn't think to rummage for a pair of braces.

And yet the first few bluebells are beginning to bloom and dip their heads. I feel impatient for their heady, almost overwhelming scent. Not long now.

It's the weekend and we're all working hard to build the new veg beds. Fifteen tonnes of compost have arrived; like a giant, erupted molehill, it sits so deep, rich and tangy, that I'd happily lay down in it.

A couple of days ago, I grabbed an umbrella and took an early morning walk with Billie into the wood and down to the small pond we dug out about ten years ago. The original hope was to create an oasis for all the local wildlife that might be passing, in search of a drink or indeed a safe haven. Over time it seems to have become a favourite watering hole for the local deer population. The edges have been churned to a bog by their frequent visits as they brace, stretch and dip to take a draught.

Sadly, it's impossible to approach the pond in 'stealth' mode with Billie as she only has one setting, and that's 'bounce'. Inevitably we've disturbed the wild mallard couple that has taken up temporary residence. They'll be back, no doubt – but they're brave to make this their home. Last year we lost all of our chickens to a very determined vixen who had cubs to feed. We know this, because we saw her, a lot, and actually watched helplessly as she trotted off with our last hen held loosely between her jaws, flailing feebly. It was quite the most sorrowful sight – one that left me feeling utterly wretched and swearing we'd never keep chickens again.

The fox's final performance was bittersweet – we spied her as she and three cubs lazily played among the long, bleached grasses in the bottom corner of the field. Any resentfulness I felt was immediately quashed by a flood of respect. She'd successfully raised a lively, healthy family. We'll just have to get a whole lot smarter if we want to keep chickens again.

Perhaps this choice of home is as good as any for those ducks, but having already found one forlorn and broken duck egg in the long grass, I fear for any future brood.

The flag irises are a little way off from budding, but the Saint Agnes' flowers are now in full bloom, their snowflake heads nodding and mirrored in the quiet waters. I can't wait for the blink-and-you'll-miss-it zoom of dragonflies that frenzy-feed later in the year; the air is warmer and so pond life peaks, with thick and whining clouds of excitable midges gathering above the zig-zagging pond skaters and rowing water boatmen.

This morning, another walk, led by the need to check on the hives and then further to the greenhouse, to water and encourage this year's seedlings. A cheerful trio of nuthatch, coal tit and dunnock stand out as the noisiest choir members, chiming in the day regardless of the fact that once more it is absolutely, relentlessly, hosing it down. Nevertheless, it's still good to be out. The rinsed air feels soft and smells sweet. Taking this walk almost daily, I've recently noticed that the old hornbeam and holly hedge that hems the field has become decisively re-punctuated by perfect, badger-shaped holes. Billie and I make our way to the gate at the west corner of the field, onto the rutted lane. Turning right, we walk back up in the direction of home to take a proper look at these new earthworks. I can see, from this side, they've made tidy work of their runs. The paths are scarped, but almost stepped with the same repetitive pattern of footfall. Wayward twigs within the tangle of branches appear flattened by muscular shoulders and wide flanks. Each entrance is banked and daubed with beguiling constellations of primrose and wood anemone that seem to glow in this soggy half-light. It's almost as if they've taken pride in their reclaimed 'doorsteps'... 'Brock Lane' has never looked so orderly and positively decorative.

When we first moved here, there were badgers – we knew this from the salt and pepper tufts of hair knotted along the old and rusted barbed wire at the bottom of the field, the well-maintained latrines dotted throughout our wood, and from the occasional torchlit sightings of dear and bumbling backsides caught in retreat as they hurriedly nosed their way off the lane, through the hedge and into the night. Then all signs of their tenancy gradually petered out and I feared that we'd not see them here again. But there's one clear 'advantage' of having dogs. And that's if there's a new brand of perfume in town, then everybody wants to be the first one to sample it. And the smell of badger poo is unforgettable.

And so the badgers are back. Controversial with the local farmers perhaps, but I'm delighted.

Summer

For the Love of Flowers

Most mornings, no matter the frown of cloud above,
I manage to execute what I think to be a finely honed
routine of creeping down stairs, avoiding the steps that
chatter back, quietly grabbing some wellies, perhaps a coat
if there's one en route, and then making my way through
the kitchen and out of the back door.

Just recently, I've realised that throughout this assault
course my stomach is clenched and my breath held. God
knows why. It's not as if anyone's going to ambush me
as I reach for the 'handle to freedom'. Perhaps it's born
of times when the children were very young, and those
same griping stairs would be enough to trigger a domino
effect of restless calls along the corridor – whereas now,
of course, I'd love nothing more than for one of these
teenagers to wake and join me in stealth mode as I slip out
into the early light.

Living on the edge of a field that remains pretty much
untouched (except for topping in late summer) means
that I have the undeniable luxury of being able to enjoy
each season as it plays out over every dip and rise of
this small patch of countryside. It's banked on the east
side by the tightly entwined arms of ancient hornbeam,
holly and hazel, held in check along the south edge by
an old millpond and cosseted on the west side by ancient

woodland that once held the delightful, earthy title of Brooker's Rough.

Today, as I step out of the kitchen door, I'm wearing my husband's dressing gown, and my daughter's old wellies (first gently shaken to encourage the exit of any overnight prisoners – beetles, spiders, the occasional wasp).Last night, as the sky flooded with indigo, the song thrush filled his chest and let fly his now regular soliloquy from his leafy perch in the old oak. This morning, however, it's the wren's turn. He usually clears his throat at around 4.30 and by 5 he's into his stride – melting my heart regardless of the mental fug of tiredness from such an early rise.

Down through the field, lifting the hem of the dressing gown, but not quite high enough; the grasses have grown and each spear is loaded with dew.

The islands of yellow rattle have spread and shifted from last year's anchorages. They're beginning to bloom yellow plumes, like miniature cockatoos perched and peering out from the rigging of a mast. Red, striped, white and a shy ramble of suckling clover are now filling in the stubby areas where the grasses haven't quite managed to take hold. These low-lying, colourful pompoms tend to flourish where the deer like to feed, also happily sprawling their Pollock-esque palette where we've mown a path to the veg plot. Hitching up the dressing gown, I strike out into the ocean of green.

It's a little early in the year for the grasshoppers, so there's no one jumping into my boots as I weave my way down through the tall grasses that whisper against thighs and tickle at outstretched palms. I'm running a mental tick list as I tromp, now wet legged…

Yellow archangel
Fox-and-cubs (I very much love this fiery-faced beauty)
Common spotted orchid
Rough comfrey
Bird's foot trefoil
Meadow buttercup
Soft red spears of young sorrel

Overhead, the quiet air is sliced by the keening cry of a buzzard that, for the last 12 years, has made this his home and, together with his mate, has successfully reared a new generation almost every year. She will be sitting on their clutch right now, though in a thoroughly modern way these two will share parental duties. A couple of days ago this buzzard received quite the dressing down from two rather rattled crows. Crows will defend any perceived threat with apparently little regard for personal safety, but buzzards are unlikely to fight back mid-air, preferring to dip and dodge the airborne assault. So the sweary acrobatics ended as quickly as they began with the buzzard ducking out and heading back to his nest.

The light blue puddles of common field speedwell are still very present – and very welcome – and as I make my way towards the guard of silver birch that edges the forest, I can see the pale mauve splashes of wood speedwell spilling out from their shady earthen beds.

It seems just yesterday that this forest was a carpet of fragrant blue, and yet it is now completely engulfed in an ocean of chest-high bracken. And it's humming! The horseflies are back. There are very few creatures on this earth that I struggle to admire, even if I don't like them. Horseflies fall into this category. I'm sure if I searched hard enough, I would find some quality that I could applaud. But I'm pretty certain that the silent, stealthy, weightless bite of a horsefly would overshadow any perceived appreciation for such a relentless and opportunistic diner.

I increase my pace, flapping my arms rather uselessly, wishing I had a few more appendages – with lasers to zap the flies. Trotting up through the narrow track banked by sandstone and outcrops of wild and flowering rhododendrons, I reach peak 'windmill' and run through the overgrown glade of yet more bracken to the safety of my studio, slamming the door.

I'm home now and I've made coffee. I'm also trying to find that tube of Savlon to rub on the red and rising bites.

A few weeks ago, my mother mentioned a book that she swears she gave me a long while back when my grandfather died. But I had no recollection of having it.

I've just found it. It's a 1965 edition of *The Concise British Flora in Colour*, by Revd William Keble Martin. Its pages are foxed and much yellowed and there are many varieties of flowers I know to be missing. But the illustrations by Keble Martin are beautiful. They're botanical, honest and clear. But all are carefully portrayed with a genuine love and enthusiasm for his subject, and as such they are radiant. Among these pages I've found scraps of paper with notes written in the hand of my grandfather: plants found, dates of discovery, the telephone number for Kew Gardens.

A pressed meadow buttercup greeted me from plate 3, and I found a common spotted orchid slightly stranded among the geraniums of plate 19.

I'd somehow forgotten that my grandfather, often an abrasive man, mercurial, and frustratingly contrary, was at his core deeply in love with nature.

Bird's-foot trefoil

Fox n Cubs, Death Paintbrush

Meadow Buttercup

I'm holding this book and I'm feeling quite tearful. I'm glad I found it, and I'm grateful to have at least a bit of that difficult old bugger nearby. He was an impossible man to navigate, let alone love. But there were many aspects of his character that I adored. Meandering through his beautiful book has occasioned quite a miracle. Like a smear of Vaseline on the edges of a camera lens, I find myself able to focus on just the best bits about him. I think that's quite a gift, and I'm sincerely grateful. To brood on the less lovable qualities of someone who's no longer around to defend themselves can be quite corrosive for those still living.

It has struck me, not for the first time, that the physical action of turning the pages of a 'real' book may grant the reader a little breathing space and latitude for the consideration and subsequent nurturing of an alternative, perhaps enriched viewpoint.

God bless the flowers, and long live the printed page.

Plum

I can't remember the first plum I ate, though I suspect it was perhaps a tinned, rehydrated prune, the kind you're offered in a pudding bowl with others, forming little shrivelled islands in a thick yellow sea of Bird's custard (which I love). I do recall the heated debates that would ensue, if either my brother or I got the 'right' number of prunes that meant we would henceforth have the indisputable *power* to predict a wonderful future. In fairness, it was blatantly geared towards girls, and the assumption that they would aspire to marry – and marry a 'rich man' at that. The list of potential suitors could perhaps do with an update, to include, among others, yoga guru, ecowarrior, hipster.

Come summertime, plums would appear in a bowl on the kitchen table, nestled amongst Cox's Orange Pippins and a few early pears. But I'd rarely reached for them, preferring the convenience of an apple that didn't leak and make my hands awkwardly sticky. And anyway, Mum's plum tart was always a winner pudding for Sunday lunch.

If only I'd known…

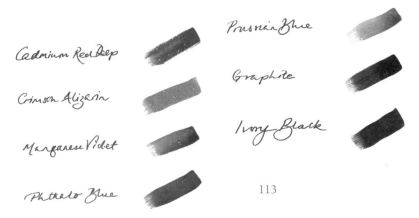

Cadmium Red Deep

Crimson Alizarin

Manganese Violet

Phthalo Blue

Prussian Blue

Graphite

Ivory Black

Now I'm a 'grown up', with a family and a bit of land, where we planted apple, pear and plum trees. They hunker down in a little orchard that we fenced off from the deer; they would strip the young trees within minutes, given the chance. In July, it is one of the greatest pleasures to slip down, alone, to the orchard, and pick a sun-ripened plum; standing beneath the little tree, with the sun on my back, a warm and voluptuous plum in my hand – the wave of delicate perfume radiating from the blush and bloom of this softly glowing fruit. A flavour bomb waiting to happen; utterly beguiling.

And then there's that first bite. The lick-of-the-lips smooth skin as tongue guides teeth to the plump 'give' point… the spare hand under the chin, as teeth cut into flesh – it's always messy. Juice will always roll down from palm to wrist, and, if unchecked, will trickle to the bony tip of an elbow as – throwing all caution to the wind – you go in for a second bite. It's perhaps one of the most sensuous fruits, second only to a ripened fig. Summer love, in all its fresh and sweet abandon. And no one wants it to end… so you reach for another plum.

As I write this, I'm working my way through the leftover 'models' from a recent project… 'tinker, tailor, soldier, sailor…' and I feel as if I've cheated a little, enwrapped as I am in this deliciously sweet moment before time… it is only May.

But it was worth it, and in a couple of months we'll be enjoying plenty of our own.

Three Square Miles

The same busy little colony, gathering from the same three square miles.

From the thermal-riding buzzard's view it would appear as just a tiny pocket in the rolling patchwork of textures and oft-repeating patterns that weave together the dips and curves of rural Sussex. But from our bees' perspective it perhaps represents their equivalent of a supermarket aisle, and, hopefully, a fully stocked one.

In the month of May, a local farmer planted a 12-acre carpet of oilseed rape, the bees' quick-fix equivalent of a McDonald's, and our crab-apple tree was also a popular meeting place in late April. Each bee met the apple blossom of her eye, becoming almost incapable of leaving the floral feast for overloading her back-leg baskets with pollen and her tummy with nectar. I would pause to watch in amazement as they carefully packed it all up and staggered along an invisible runway to take off and head home, pushing the weight limit of carry-on luggage to the extreme.

The field edges began to fill with cow parsley, and the occasional cowslip, while the hedges became fringed with white clouds of blackthorn.

The rich, sweet honey they made has set like concrete, and wherever it's spread it stays put, resolutely immovable

May ~
Raw Sienna

July ~
Yellow Ochre Pale

June ~
Burnt Sienna

until in the mouth, where it melts with the grainy texture of tablet fudge.

As May slipped into the blowsy pocket of June, pink pillows of field mallow called out to passing bees to perch, rest awhile and feast. In the vegetable garden, pea and broad bean plants offered up bounty that lay barely hidden among the pastel folds, and the nearby dappled cool of forest paths became lined with parades of trumpeting foxgloves. But the overwhelming flood of temptation came from the huge outcrop of broom that offered up its sunny flanks at the top of the field. It became such a go-to place that eventually both colonies made this their restaurant *de jour* and a daily string of bees would rope their way to and fro, across the field. We learned to duck and dive beneath this visible highway, marvelling at their enthusiasm and ducking the drunken bees' occasional collision with an unaware human.

The honey they conjured tastes like a holiday in the tropics, a gentle sea breeze, the soft, fleshy fruit of a cocoa pod laid open, almost toffee…

Then came July and the honey turned the colour of straw. With the field now a cheerful canvas from summer's palette, walking through was no longer a quiet affair. Competing with the raspy chirping of grasshoppers, the bees could be heard happy-humming from all corners,

with white and pink clover and bird's foot trefoil being
the most popular hangouts. In the vegetable garden, late-
flowering courgettes and pumpkins enticed scouting bees
to change their dance on returning to the hive, and the
hedges became rammed with bramble flowers.

July's honey pours like unpasteurised cream, and smells
and tastes like a summer pudding – intensely, deeply floral.

Each jar is like a window into another time, holding
the unadulterated essence of a month of flowers. And, of
course, these aren't calendar months – because flowers
don't dance to our tune and our need for organising
and scheduling every minute, hour, day and month of
the year. They don't bloom for us, they bloom for the
bees, gardeners in the truest sense. For the last five years
I've been lucky enough to observe this extraordinary and
intimate relationship close-up. And the more I see the more
I realise how fluid this alliance is, but how vulnerable too.

I find it hard to believe just how much our rhythm of
cutting, pruning, sowing, protecting and harvesting has
changed. Every activity used to be geared entirely towards
feeding my family; now it's done with the bees' welfare in
mind: what would make the bees – and by extension all
pollinators – happy.

I suppose taking a little of their honey could be seen as
just rewards for ensuring they've a full and varied aisle to
gather from. But in truth it still feels a little like robbery.
As such we only take enough to enjoy for ourselves,
and if there's any spare I give it away. I'm often asked if I

sell my honey to 'the public'. I don't think I could. I feel strongly that honey should be seen as unique, a luxury of inestimable value. The minute it has a price tag, it makes the work of the bees a commodity. That wouldn't be a bad thing if they were rated as highly as oil or gas but, sadly, for all the newsworthy rallying that goes on in efforts to 'save the bee' we're still a long way from that. I would rather it didn't have a commercial value. But to give it to family, friends and sometimes complete strangers, feels good. There's the lady with hay fever at the Post Office who'd read that local honey might help ease her symptoms, the phlebotomist who patiently waited while I did star jumps to get the blood pumping so my veins would actually rise to the occasion. An opera singer who was on the verge of losing her voice. And the friend whose three little boys value their pots as we might, if in possession of the winning lottery ticket. And the dear friend who grows things to find peace and joy and understands their rhythm. The list grows longer each year, and I'm reaching towards the end of this year with far less for us. But there's enough to see us through winter. These moments sit well in my heart.

As August gathers pace and summer begins to fade, the bees are now gathering up the last of the nectar and pollen to store in their larders and sustain them through the cooler, darker months. The ivy is proving to be irresistible. They've been clamouring around the stubby explosions of nectar lollipops like kids in the sweet shop after school. As I take my habitual walk around the field – and beyond – I can see that the three square miles of foraging is looking a little empty, like a supermarket after the bank holiday shopping spree.

But the flowers have been pollinated and, with a fair winter, come spring, the aisles will begin to fill once more.

August in the Orchard

With barely a whisper of breeze, July has slipped her
mooring lines and, lifted by one of the zillion thermals
that this extraordinarily warm summer has conjured,
she's sailed away with barely a backward glance as August
rummages for her bathing suit and broad-brimmed hat.

To be honest, it's been a very long time since the month
of August required a sun hat, let alone a bikini. Usually,
most of us are reaching for our wellies and raincoats as
we contemplate the British summer holiday we sensibly
booked a year in advance. But this summer has continued
to stretch herself out and languish just a little while longer
on her beach towel.

And, of course, I've now jinxed it, haven't I. So you
know where to find me if it all goes horribly wrong. I'll be
standing in the cloakroom ready to hand you your boots,
coats and soggy pasties.

But for now, it's still hot, and there's no sign of rain for
the time being.

Yesterday I slipped on my boots and took a walk down
through the field to check on the fruit trees. I could see the
heat rising in a shimmy of waves from the still, dry grasses,

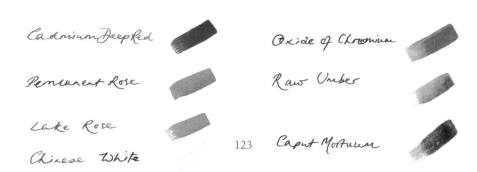

Cadmium Deep Red

Permanent Rose

Lake Rose
Chinese White

Oxide of Chromium

Raw Umber

Caput Mortuum

and all around me the parched symphony of crickets and grasshoppers vibrated loudly from the taller, bleached islands in this thirsty ocean. These string musicians seem fatter than previous years – their legs more elastic, their bounce more ambitious. Wearing boots is, of course, a wise thing to do, with a view to possible grass snakes, adders and ticks – they're also usually high enough to avoid the odd grasshopper jumping in. But apparently not this year! I must've stopped half a dozen times to wobble, one-footed, as I shook out yet another unfortunate passenger.

With all this sunshine it seems that the apples have decided to ripen a little earlier. They're somewhat smaller than their usual comforting handful, and I would think that's a direct result of the lack of rainfall for the last couple of months.

With no basket, I was ready to pick just a few and pop them in a hastily gathered T-shirt pouch but then I gradually became aware that I had company. Tuning out the hiss and rattle from the 'hoppers and the husky refrain from a trio of tenor pigeons, I slowly switched frequencies to pick up the unmistakable deep grumble from a busy squadron of hornets. It seemed I was supposed to reserve a table at this particular restaurant. I decided to leave them to their endless taster menu and headed back home.

Egged on by terrifying stories (possibly embellished to augment the sense of terror) of violent wasp attacks and killer hornets, most of us tend to be at the very least quite wary of anything with a yellow-and-black stripe. There's something instinctively alarming about the hard-edged, striking pattern created by these particular colours that evokes a sense of foreboding in us all. And I'm sure there was once a time when such an encounter with any stripy-rumped insect would have left me frozen or running for my life. But the truth is I can't remember it!

What I can remember is at the age of about nine, sitting at my parents' kitchen table, an open pot of Mum's strawberry jam nearby, and my father placing a small dollop of the jam on my outstretched finger. It was a hot summer's morning and we were in the midst of breakfast. A wasp had entered through the open window and had spent much of its time dipping down indecisively between the jar, the lid and the open pack of butter.

Dad told me to hold my finger out and stay as still as possible. Blind trust can be a complete disaster sometimes, but it can also be an amazingly liberating state of mind. And as a nine-year-old child, I trusted him implicitly. So I sat quietly, leaning forward, my legs wound around each other for comfort. The wasp briefly hovered above my finger, then left. It circled a few more times, almost as if it couldn't quite believe the proffered banquet.

And then it swung back in and gently alighted next to the splodge of jam. I could feel its feet!

Very slowly, it began to work its way along the edges of this sugary puddle. Antennae waggling, head down, it appeared to be actually sniffing the jam. Then the wasp began to lick, and it kept on licking, all the while delivering a nuzzling, micro-tickle to the tip of my finger. At one point it paused to wipe down its antennae with its front legs – as one might clean off a windscreen wiper that's got a little grubby with road dust – and then carried on licking, tail-end bobbing, perhaps with enthusiasm. Astonished, I began to giggle. I can still remember the delight.

As a child, I felt this to be an extraordinary discovery that needed to be shared, and it became my 'party piece' when adults popped round for a drink, and I endeavoured to delight my friends with the same thrill, if I could just get them to stop flapping their arms.

Looking back it was perhaps one of the most important lessons I learnt as a child. And I've tried to do the same for our three children.

Keen to collect apples, but with this slightly sugar-crazed gathering of hornets, bumping around the branches of the tree, I decided to pop my beekeeping suit on, legs rolled up for fear of spontaneous combustion. I returned to the orchard and, working carefully with extendable pruning shears, I managed to duck and dive among the branches, avoiding the rows of striped bottoms that lined the empty hollows of apple husks. There's plenty for us all this year, so I didn't feel it necessary to argue for the odd apple. I plucked and caught a full basketload and left the hornets to their feasting.

As the day continued, clouds gathered and the air became heavier. The kitchen windows were pushed as

wide as they would go. Looking through the apples that
I collected, I could see that some of them weren't quite
ripe, while others bore the tell-tale scars of the hornets'
mandibles. I decided to slice and freeze a few pounds for
winter crumbles, make compote with some of the less
sweet ones for the children's breakfasts and put aside a
generous apronful for an apple tart.

 As I began the meditative task of peeling and slicing,
I was joined by a lone wasp, no doubt hunting for
something sweet. I put my knife down and reached for
a pot of January's marmalade. Opening it up, I dipped a
finger in, then held it out, and waited.

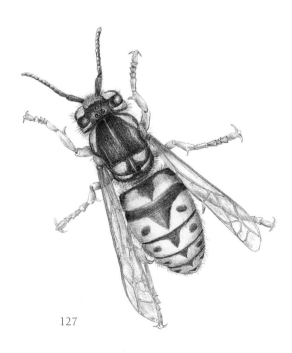

Pears

If memories could be stored in boxes, this summer would have filled the multitude that still sit in our attic from our move here, more than 19 years ago. Except memories should never be allowed to be stored, sealed, and mothballed. This summer, I have never felt more keenly the need to keep making fresh memories and I have watched and mentally 'clicked' on as many moments as I possibly can fit in my head.

About ten years ago, we made the decision to close up the tiny veg beds we'd hastily dug and planted when we first arrived. The children were growing in direct correlation to their appetites, and these infant beds had become too small to provide the rocketing volume of vegetables needed to fill five tummies. We had the added challenge that these little earthy squares had become increasingly seasonally challenged by the growing cloak of shadow bestowed upon them by a flourishing oak that we rather naively hadn't factored into our plans. And while veg beds are movable, old oaks trees are not. So we took a walk through our field and hatched a plan for a new, larger and more open plot. We dug and fenced an area the size of a tennis court. Dividing it into three, the central section became our grown-up veg plot, and with the

Oxide of Chromium

Cadmium Yellow

Caput Mortuum

Sepia

Ultramarine

Venetian Red

Titanium White

children's help we laid down bricks and built raised beds.
With another third we were excited to be able to plant up
soft fruit beds, and a long, thin asparagus plot. The final
slice we devoted entirely to fruit trees. We may be ball and
chained to a heavy mortgage, but we felt truly rich with the
promise of such bounty. At the time, it was seen as essential
to build this big if we were to keep up with our hungry
brood. But in truth, and with glorious hindsight, it was
perhaps a little ambitious, given the sheer amount of time
and effort that it needs to feel even vaguely in control of
this jungle.

When we first built these new beds our three children
were small, yet very keen to be involved. One loved
carrots, and we found seeds promising many colours and
created a Union Jack of sorts. Another wanted an entire
bed of parsnips to roast. Our youngest was obsessed with
pumpkins and so together we grew some monsters. Many
images will be forever imprinted on my heart: dimpled
hands wielding small tools to rootle out weeds; hand-me-
down sun hats shielding eyes as the children staggered
with watering cans to feed their little families. Our middle
child decided that snails should no longer be jettisoned
over the proverbial (and literal) hedge, but must be
given new homes in which to see out their days with as
much 'spare' green and leafy stuff as they could possibly
consume. Far be it from me to explain my particularly
hardened view towards these freeloading feasters to my

kind-hearted children, so I found some spare bricks to help build this new hotel!

As the years passed, the time our three had available to continue growing their food (and feeding it to the snails) diminished rapidly, to be replaced with school, increasing quantities of homework, and a social life beyond the curtilage of our home and field. And so, naturally, they weren't quite as able or keen to continue. At this point I could get rather morose, but of course I'd be a complete hypocrite to start ploughing that particular field, since I can remember doing exactly the same with regard to my parents and their green pursuits! And so, inevitably, the vegetable garden became more 'mine', not because I liked it that way, but simply because the digging and decision-making was more often than not left to me.

With them all now firmly entrenched in teenager-dom and beyond, they're less involved than they ever were, and this tennis-court-sized space can occasionally feel a little overwhelming. But now, more than just a vegetable garden, it has become the perfect destination when seeking a little order and calm that can't be found elsewhere.

Meanwhile, I console myself in the firm belief that each of our three offspring will, at some point in their lives, find themselves returning to the soil, just like I did, when they once swore they'd be as likely to pick up a spade as a snake. The easily found comfort of planting and tending seeds has a habit of slipping back into the frame, usually unnoticed, and I daren't tell them that it often goes hand in

Must net pears from wasps & hornets.

131

hand with listening to the radio and talking to the blackbird at first light, both of which I'm happy to admit I do.

Throughout my pursuit of growing our food, I've been afforded the freedom to grow whatever I like, regardless of whether I'm actually going to eat it, simply because I have the genuine excuse that I'll be illustrating it. I'm growing my models. I run a modelling agency for the green, the good, the weird and wonderful. My choice to specialise in food illustration was founded on sticking with something that *meant* something. And when we created this overly ambitious plot it was a blessing, and it massively enriched my opportunity to portray edibles in their most alluring state of freshness.

Those who grow their own produce will know that of all the seasons, spring always seems to be most labour-intensive among the veg beds (dig, sow, plant, water, weed, and repeat) and more often than not I forget to look over my shoulder and see what the fruit trees are doing.

As spring slips into summer, the variety of veg garden chores diminishes to a select few: weed, water, re-tie the wayward. It's very easy to stand with a hose in one hand and a cup of tea in the other and get wholly lost in thought while the grid of onions soaks up its allotted slot in the daily watering regime. Just occasionally, I find myself slightly resentful of this chore and rue the daft decision we made to plant quite so much. But mostly I relish this time of the day for being markedly calmer and less demanding than the hours spent hustling and bustling my busy children. More often than not there's the gentlest sense of being watched. It's rare that he'll interrupt, but I know it's the blackbird who likes to sit in the old holly bush and keep his sun-rimmed eye on me and whatever I might have unearthed from the soil.

By now in the orchard the swathes of candy-floss blossom have gone. The flowers have wilted and their petals have pooled like piles of discarded ticker tape beneath our young trees. And it never ceases to feel quite the miracle, to find tiny embryonic fruit have begun to emerge. The apple trees are always the first to brandish their wares – as pea-sized as they are. The plum trees are next, clusters of hard little olive-like bullets that will rapidly swell and overload their arms once more.

Most mornings, I'll head down in my dressing gown and boots to check on the progress of everything, often with a mug of tea, always with a butterfly churn of anticipation.

One such morning late this summer, I made my way around my favourite apple tree (don't tell the others) and while my hand found its way along one of her arms to a truss of ripening fruit, my eye was caught by the sight of large, pendulous droplets of green on her neighbouring pear tree.

It seems that for the first year ever, both of our young pear trees have decided to push on past the bloom, and braving ridicule from the more productive apple and plum, they've produced a total of nine Williams, and eight Conferences. And I almost missed them. Yet there they are, dangling enticingly among the curtain of shiny, ovate leaves, 'almost' within my grasp – and definitely within a beak's bite of a keen-eyed bird.

But now I know where they are I've been watching them closely, occasionally jumping to give the lowest Conference a brief and tentative squeeze.

The second it yields under the pressure of a thumb, I'll fetch the ladder.

Straight from the tree, there are really only two ways to enjoy a pear. You can cup its generous bottom, and, with the briefest of crunches, you're straight through and into the flesh – or you can choose the more considered and elegant route of plate and knife. I have a penknife that I found while climbing in France. It was among the dust and rockery of a natural ledge that formed part of a steep ascent, and I chose to pause there awhile and take a sip from my flask of water. A sad loss for its original owner, it's a beautiful old thing – simple, elegant and with just one blade. To pierce the skin at the tip, slip the blade down as the pear widens to its rump, and open out a perfect twin of creamy white, is a joy… and just as messy. That sweet burst of perfumed juice, followed by an unconscious knuckle-wipe of a wet chin.

Throughout history pears have been both revered and ridiculed for their shape, lack of flavour and gritty quality. Shakespeare referred to them as either 'absurd or unpleasant'. It's clear that he never experienced the heady bliss of a perfect pear.

Ripened on a tree, (or a windowsill, if the birds are already queueing along the branches) it is perhaps the most sublime fruit to eat as is, unfluffed or adulterated with any pastry or pomp.

On spotting these little green miracles, I have considered grabbing the ladder and trying to find a way of netting them from the birds, but then realised that I'd still be too short and would no doubt end up wearing the net.

Instead, I'll bide my time and keep an eye on these pears and hope that I can beat the crows and blackbirds.

I'm now sitting in my little studio, and I'm finishing a painting of some Conference pears, in egg tempera. And, of course, it's difficult not to draw a parallel with our dear children, who, like the pear trees, have blossomed and now fruited into young adults – and I almost missed it.

So this summer I've been stashing away these precious moments, to the point where there's more than a slim chance that I could just burst. There's a driving sense of urgency that I feel, like trying to stuff the feathers back into a split pillowcase, before they float off, out of my reach.

It's unlikely that we'll ever get around to unpacking those boxes in the attic. The contents are perhaps not that important, or surely the taped lids would've been torn open a long time ago. But consciously stockpiling these memories feels like a vital act. I'm certain I would regret it, if in quieter moments, I wasn't be able to reach in, take them out, air them and rekindle the warmth of each moment as it was. Time to go and check on the pears.

Days Like These

The gentle slips and slides of one month into another are often punctuated with some of the most unexpected and breathtaking displays of a season in flux. Curtains drawn back on an early morning can reveal a stage filled to flowing with a swirling murk, sifting the day's first rays that eventually manage to push through to backlight dew-laden, summer-weary growth, rendering everything it touches gilded and vivacious once more. Indeed, some of these performances deserve a standing ovation.

Summer has focused her gaze on the sun's shifting arc and now, with an easy and remembered rhythm, is steadily making her way towards autumn with all the confidence of a seasoned pro.

And yet there's still a warmth in the air. When the sun's light reaches these low-slung mists, they puff into embryonic plumes that become liberated and lifted to surf among the silent gathering of fledgling thermals. Days that begin this way are destined to be labelled as halcyon, bucolic. Less Turner, more Titian. I try hard to commit the fugitive moments to memory, but always with the hope that it won't be needed as there'll be a repeat performance showing soon.

As ever, most of these early mornings are spent jamming bare feet into wellies and retying my dressing gown to head out and into the field for a walk, look and listen.

More often than not I return with a wet hem and an armful of something that makes me happy. With the kindest of Septembers, cheerful dahlias, tomatoes and figs have featured frequently in these last few weeks.

But there's nothing quite like seeing a morning through someone else's eyes.

And so, a few days ago I set my alarm to an ungodly hour that no bird would consider sensible and drove through these mists to meet a good friend on the edge of a little village that sits within one of the many dips of the South Downs. With hushed words (I'm never quite sure why we do this, but we do, don't we… as if we might wake up the birds – the earth even) we made our way through the meanders and forks of the rutted tracks that flow through these well-trodden ways. These paths were banked by crowds of rusted, long-limbed wild carrot, empty baskets held high to sway with the moving mist like a sepia sea. Among their feet grew a smattering of field scabious, small daubs of mauve to brighten this desolate view. We kept walking, talking, pausing to look beyond the fog, and eventually climbed up onto a shoulder of this rolling body of land.

As we stood just above the quilt of fog, my mouth became a silent 'O'. So many skylarks! No longer the joyful ascension to babble and bubble unseen, but now in plain sight. Their song, delightfully conversational as ever, had taken on an air of busy urgency as they jousted and swerved, engrossed in the game of territorial rights.

Having heard but never seen a skylark before, it seemed that we were now the invisible ones. I stood so close to them as they rode the sky. I whooped and wowed as they scored and sliced through the fog-heavy air, courageous riders of some indiscernible rollercoaster.

To look at the bird with an objective eye you would see that there's nothing particularly impactful in their attire. Skylarks are of medium size, with a medium length of wingspan. Their markings, being muted flecks and bars of buff, biscuit and cream, are designed for camouflage among the cropping fields in summer and beyond. The most distinguishing feature might be considered the quirky brown-streaked crest, which can look fairly punkish when raised. But to put it bluntly, they're not exactly catwalk material in the bird world.

Yet to actually see them at all felt wholly miraculous. And so, of course, to me, they were all the more beguiling for their unassuming choice of plumage. My breath was taken and replaced with a smile as wide as our view.

As I gathered myself together, filing away this extraordinary memory under 'unlikely to see again', we made our way back down through the cool pockets of lingering mist and into the nearby town of Lewes. While my friend headed on foot to his office, I caught a train back to the little village where I'd left my car.

I'm sitting here, writing about this moment and yet when I close my eyes I have visuals of other times throughout my childhood, memories of my father and his birds. And

I'm almost flooded to overflowing with the technicolour, surround-sound joy of them all as these images flicker past like a zoetrope on full tilt. Eyes open and I'm aware that the morning is still young, yet the sun has managed to elbow her way through the fog and I can see our field beginning to glow.

It's time to stop writing now, make more tea, and head up to the studio. There are pigments to grind and a Porthilly oyster waiting to be painted.

Thank You

Koska, thank you, my love, for your unfailing support, always at the moment when I felt sure I'd messed up. You even braved a potential storm and dared to point out where things could be improved. And you were right, every time.

Flora, Ellie and Jolly, my beloved cheerleaders, thank you so much for your unalloyed enthusiasm throughout.

Anja, I'm so grateful to you for your insistence that the time was right to make this book. If it hadn't been for you, I'm certain I would still be screwing up balls of paper.

Emily Sweet, my agent, thank you for your kindness, optimism and patience, and your willingness to take me on.

To you all at Pavilion. What a formidable team! You've taken my idea and made it so much more beautiful than I ever could've hoped it would be. It's been a joy to work with you. Thank you.

Finally, Mum and Dad, who moulded me this way. I am beyond lucky to be one of the acorns that fell from your arms.